Recent Advances in Carbon/Graphite Coatings

Editor

Gianfranco Carotenuto

MDPI • Basel • Beijing • Wuhan • Barcelona • Belgrade • Manchester • Tokyo • Cluj • Tianjin

Editor
Gianfranco Carotenuto
Institute for Polymers, Composites and Biomaterials—National Research Council (IPCB-CNR)
Italy

Editorial Office
MDPI
St. Alban-Anlage 66
4052 Basel, Switzerland

This is a reprint of articles from the Special Issue published online in the open access journal *Coatings* (ISSN 2079-6412) (available at: https://www.mdpi.com/journal/coatings/special_issues/Carbon_Graph_Coat).

For citation purposes, cite each article independently as indicated on the article page online and as indicated below:

LastName, A.A.; LastName, B.B.; LastName, C.C. Article Title. *Journal Name* **Year**, *Volume Number*, Page Range.

ISBN 978-3-0365-5725-0 (Hbk)
ISBN 978-3-0365-5726-7 (PDF)

Contents

About the Editor

Gianfranco Carotenuto

Gianfranco Carotenuto is Research Director at the Institute of Polymers, Composites and Biomaterials (IPCB) at the Italian National Research Council in Naples. He received his degree is in Chemistry (1991) and PhD in Materials Technology and Industrial Plants (1994) at the University 'Federico II' of Naples, Italy. He was a post-doctoral student (1995) at The Center for Advanced Materials Processing (CAMP), Clarkson University, Potsdam (New York State, U.S.A.); visiting scientist (1996) at INM (Leibniz-Institute für Neue Materialen) in Saarbrücken (Germany); post-doctoral student (1997–1998) at the Department of Material Science and Production, University 'Federico II' of Naples; contract researcher (1999–2000) at the Institute for the Technology of Composite Materials (ITMC-CNR) in Naples; and then a researcher (2001) at the Institute for Composites and Biomedical Materials (IMCB-CNR) in Naples. Dr. Carotenuto became a contract Professor (2000–2010) at the Engineering Faculty and the Letters and Philosophy Faculty, University 'Federico II' of Naples. He is author or coauthor of numerous publications on international journals in the field of material science and engineering and Editor of two books on 'Metal-Polymer Nanocomposites' published by Wiley. He has 10 international patents and has given many presentations at national and international conferences. His Scopus H-Index is 23 (as of 8 January 2020). His present research interests are the following: functional microcomposite materials, metal–polymer nanocomposites, solid-state ionic conductors, polymer-supported graphite nano-coatings, polymer-embedded nanostructures, chemical sensors, and sustainability.

Preface to "Recent Advances in Carbon/Graphite Coatings"

One of graphene's greatest potentials is to reverse the physical/chemical properties of the surface to which it is applied. Surfaces treated by graphene switch from dielectric to electrically conductive, from hydrophilic to hydrophobic, from opaque to highly reflective, and so on. Graphene coatings are optically transparent layers only a few Angstroms thick; however, they are very effective in reversing the physical/chemical properties of a substrate. The purpose of this Special Issue is to collect some selected examples of this important technological potentiality of graphene and graphene-based coatings.

Gianfranco Carotenuto

Editor

Editorial

The Engineerization of Physical and Chemical Phenomena

Gianfranco Carotenuto

Institute for Polymers, Composites and Biomaterials (IPCB-CNR), National Research Council, Piazzale E. Fermi, 1, 80055 Portici, NA, Italy; gianfr.carotenuto@virgilio.it

Citation: Carotenuto, G. The Engineerization of Physical and Chemical Phenomena. *Coatings* **2022**, *12*, 1282. https://doi.org/10.3390/coatings12091282

Received: 31 August 2022
Accepted: 31 August 2022
Published: 2 September 2022

Publisher's Note: MDPI stays neutral with regard to jurisdictional claims in published maps and institutional affiliations.

There are a number of concepts in materials science that are only ambiguously described in the literature and/or are not completely understood. A logical analysis of these concepts may allow a better understanding, correct dissemination, and a progression in knowledge in this area of science. This sort of in-depth philosophical analysis can be very easily approached and potentially applied to a number of scientific concepts, thus becoming a novel important tool for the progression of knowledge in this scientific field. For instance, let us consider the possibility of developing a definition of functional materials that is an alternative to that currently used definition in the literature. It is not easy to provide a general and strict definition for these types of materials. Materials are typically distinct in terms of structural and functional types. Structural materials are comprehensively defined as materials capable of bearing loads. In contrast, accurately defining a functional material class is much harder than distinguishing the structural counterpart. Overtime, different definitions have been formulated for functional materials [1]; for example, they have been defined as: "those materials able to respond to magnetic, optical or electrical stimuli". However, such a form of definition can be perfectly adopted for sensing devices, but it is quite limiting for a more general case. In a device, the functional material is the "active material" that imparts the functional behavior to the device as a whole, while the exact response of the device is determined by the structure and how the functional material is integrated with the rest of the structure. Functional materials have been also defined as "those materials which possess particular native properties and functions of their own". Again, they have been referred to as: "materials that can undergo controlled transformation through physical interactions", or also as: "materials whose 'function' is associated with their electric, magnetic and/or optical properties". Surprisingly, there are also definitions based on what they are not, such as a "material which is not primarily used for its mechanical properties, but for other properties such as physical or chemical". As observed, these definitions are not really general and the formulation of an alternative definition for the functional material that is more satisfactory, clear, and universal is pivotal.

The word "engineerization" is currently used in the literature, and it can be usefully adapted to develop an alternative definition for the functional material class. Although this word has been coined in chemistry and in chemical engineering to indicate the industrial scale-up of a chemical or biochemical process (that is, the conversion of a laboratory-scale chemical or biochemical process to the pilot scale and finally to the production scale), the same word can be extended to all physical and chemical phenomena, comprising a meaning of industrially (technologically) exploiting these processes. With this preface, a novel definition of functional materials can be proposed. Functional materials are technologically useful materials based on the engineerization of some physical or chemical phenomenon. The engineerization of all physical/chemical phenomena is potentially possible for producing a functional material. This novel definition can be clarified by a few examples. Among many chemical processes, let us consider, for example, the ability of zeolites to react with gaseous acid. This reaction is known as deallumination. Therefore, zeolites nano-powders can be used as irreversible molecular traps for gaseous acids such as acetic acids [2] because of their high surface development and the ability to chemically react with acid molecules. Nano-sized zeolites powders represent a functional material of natural origin since it allows the simultaneous engineerization of the following phenomena: (i) deallumination reaction, (ii) high surface development, and (iii) fast adsorption kinetics (due to the small size of grains). A further example could be based on magnetite: This mineral is not a functional material, but it corresponds to a magnetic substance. In contrast, a magnetite-polymer microcomposite (dispersion of micrometric magnetite powder in a polymeric matrix) is a functional material (plastic magnet) since it is based on the engineerization of the magnetism phenomenon characterizing the filler. Functional materials do not identify with the physical (or chemical) property but with the engineerization of that physical (or chemical) property, which is the fabrication of a hybrid material, and its working principle is based on that physical phenomena. A magnetic and transparent magnetite-polymer nanocomposite is a multifunctional material because it results in both optically transparent and magnetic properties. In particular, when the same functional material is adequate for more than one application, it can be defined as "multifunctional" since it has different characteristics that are potentially useful for technological applications. In this case, more than one physical–chemical phenomenon undergoes engineerization in the same material. Multifunctional materials are usually considered as superior to unifunctional materials. However, is this idea really correct? A multifunctional material should be a material optimized at the same time for two, three, or more properties. Can a single material be really the optimal one for more than one application? The high specialization of materials, which is desirable for the industrial applications, clashes with the concept of multifunctionality. Too many compromises are needed in multifunctionality; consequently, unifunctionality should be better than multifunctionality. The availability of selective sensors, monochromatic light sources, specific rapid tests, etc., prove that unifunctionality should be a fundamental factor for most high-performance materials.

It must be pointed out that this definition developed for the functional material class can also be applied to the structural material class. In fact, it is possible to say that all technologically useful materials are based on the engineerization of certain phenomena that, when they are of a mechanical type (e.g., high tensile resistance, excellent surface hardness, high toughness, high wear resistance, etc.), provide origins to a structural material; on the other hand, if they are of a physical or chemical type, they should produce a functional material. For example, nanostructures, such as carbon nanotubes (CNTs), can be technologically used both for their mechanical and physical properties. Depending on which property is involved in the engineerization process, a structural or a functional material results. If, for example, engineerization involves the high mechanical resistance of carbon nanotubes, similarly to the case of CNT-polymer composites, a structural material is achieved.

Functional composite materials (FCMs) represent a very powerful solution for many modern technological problems. The most recently developed functional micro- and nano-composites, based on the incorporation of nanostructures and advanced artificial materials in a powdered form such as graphene, carbon nanotubes, fullerenes, metal nanoparticles, perovskites, etc., in matrices of different nature (i.e., ceramics, polymers, metals) belong to this important material class. Currently, an increasingly large number of scholars are involved in the development of new types of FCMs; consequently, the development of this special material class constitutes a very broad and fruitful field of research. FCMs are complex multicomponent solid systems. The different components are both molecular and solid (eventually nano-sized) in nature. The functional materials contain a structural component, usually consisting of a polymer (thermoplastic or thermosetting resin), that has the function of making a self-supporting and easy to handle material. One or more molecular or solid (particulate) components with precise functionality are also required to allow this complex system to work. Typically, a few phenomena are exploited together to produce useful FCMs; however, it is possible to take simultaneous advantage of a number of phenomena in the working mechanism of FCMs (in some cases, the functional properties are achieved synergistically by combining two or more components). Let us consider some examples of how a functional material is achieved by combining several physical phenomena. In the preparation of an optical filter or an UV-limiter, at least three physical properties need to be combined together: (i) perfect optical transparency (i.e., an amorphous polymer matrix that does not produce light-scattering phenomena because of an absence of crystallites); (ii) high solubility for the functionalizing organic molecule (dye) (i.e., in order to have high solubility of the functionalizing dye, non-bonding interactions acting among the polymeric macromolecules must be similar to those acting among the organic molecules; in addition, when organic compounds are molecularly dispersed in the polymeric matrix, they absorb independently each other and the required transparency is maintained); (iii) strong optical absorption (i.e., the organic molecule must contain a chromophoric group characterized by a high extinction coefficient value). In the case of color filters based on the surface plasmon resonance of noble metal nanoparticles (e.g., gold, silver, etc.), high dispersibility replaces the solubility requirement. All cited properties are strictly required and the lack of even only one of these properties surely compromises the operation of such a device. A further example is represented by a fluorescent varnish. In this case, the required properties are as follows: filmability, adhesiveness, optical transparency, high solubility of the fluorescent dye in the optical-grade matrix, and high quantum efficiency of the fluorescent group. FCMs can be considered as a type of molecular machine that is highly specialized in the execution of special operations. These "molecular machines" have a working principle (the phenomenon on which they are based) that can be theoretically modelled. Therefore, an FCM design is an attractive and very important area of research in materials science that is strictly required for the optimization of these useful materials.

The paper collected in this Special Issue titled "Recent Advances in Carbon/Graphite Coatings" describes the development and the characterization of functional polymeric microcomposites, functional materials of natural origin, functional materials of different molecular types, and other types of carbon-based functional materials. As observed, the developed general definition of functional materials can be conveniently applied to each one of these cases. A first example is represented by the article titled "Electrical method for the in vivo testing of exhalation sensors based on natural clinoptilolite". In this paper, a sensing device for breath detection and measurement has been fabricated by using a nature-made functional material, which is the zeolite mineral named clinoptilolite. In this case, the engineerization of three physical phenomena was synergistically exploited: (i) the physical adsorption of water molecules on the clinoptilolite extra-framework cations by ion–dipole interactions; (ii) the possibility for the extra-framework cations to act as charge carriers in an electrical transport mechanism based on cation hopping; and (iii) the increase in the charge carrier's mobility as a consequence of the water molecule adsorption on cations. A second example of functional material is described in the article, titled: "Functional polymeric coatings for CsI(Tl) scintillators", which presents a polymeric microcomposite coating developed by embedding a high reflective white powder in an epoxy matrix. In this case, the functional material is based on the engineerization of light-scattering phenomena with some micron-sized barium sulfate ($BaSO_4$) particles or Teflon (PTFE) particles that are generated in the visible spectral range. Again, an electrically conductive coating has been studied in the article titled "Influence of the thermomechanical characteristics of low-density polyethylene substrates on the thermoresistive properties of graphite nanoplatelet coatings". In this case, the electrically conductive coating is a functional material based on the engineerization of the electrical transport phenomenon characterizing the percolative structure of a graphite–polymer microcomposite deposited on a polyethylene substrate. Analogously, the graphite oxide (GO) paper coating, for which its preparation is described in the article titled "Green solid-state chemical reduction of graphene oxide supported on a paper substrate", is a type of functional material. This functional material is based on the engineerization of the physical phenomenon of the electrical transport in graphene-containing structures supported on a dielectric substrate. Finally, the developed definition can be applied to all types of functional materials described in the articles collected in this Special Issue, although they have completely different natures, thus proving the absolute generality of the developed considerations.

In conclusion, an alternative universal definition for the functional material class has been developed. This definition is based on the word "engineerization", and its meaning involves scaling up a chemical process or, equivalently, industrially exploiting a chemical phenomenon. Functional materials are materials based on the engineerization of a physical or chemical phenomenon. However, all technologically useful materials (both of functional and structural type) can be similarly defined as "materials based on the engineerization of one or more physical, chemical or mechanical phenomena", with the intention of technologically exploiting these phenomena. Such logical analysis, corresponding to a sort of in-depth rational (philosophical) study of still confused concepts in material science, could be an important tool for the progression of knowledge in this field.

Conflicts of Interest: The author declares no conflict of interest.

References

1. Vilarinho, P.M. Functional Materials: Properties, Processing and Applications. In *Scanning Probe Microscopy: Characterization, Nanofabrication and Device Application of Functional Materials*; Kluwer Academic Publishers: Dordrecht, The Netherlands, 2005; pp. 3–33.
2. Cruz, A.J.; Pires, J.; Carvalho, A.P.; Carvalho, M. Adsorption of acetic acid by activated carbons, zeolites, and other adsorbent materials related with the preventive conservation of lead objects in museum showcases. *J. Chem. Eng. Data* **2004**, *49*, 725–731. [CrossRef]

 *coatings*

Article

Temperature Dependence of Electrical Resistance in Graphite Films Deposited on Glass and Low-Density Polyethylene by Spray Technology

Angela Longo [1], Antonio Di Bartolomeo [2,3], Enver Faella [2,3], Aniello Pelella [2,3], Filippo Giubileo [3], Andrea Sorrentino [1], Mariano Palomba [1,*], Gianfranco Carotenuto [1], Gianni Barucca [4], Alberto Tagliaferro [5] and Ubaldo Coscia [6,*]

[1] Institute for Polymers, Composites and Biomaterials—National Research Council (IPCB—CNR), SS Napoli/Portici, Piazzale Enrico Fermi, 1-80055 Portici, Italy
[2] Department of Physics "E. R. Caianello", University of Salerno, Via Giovanni Paolo II, 132, 84084 Fisciano, Italy
[3] Superconducting and Other Innovative Materials and Devices Institute—National Research Council (SPIN-CNR), Via Giovanni Paolo II, 132, 84084 Fisciano, Italy
[4] Department SIMAU, Polytechnic University of Marche, Via Brecce Bianche, 1-60131 Ancona, Italy
[5] Department of Applied Science and Technology, Politecnico di Torino, Corso Duca degli Abruzzi, 24, 10129 Torino, Italy
[6] Department of Physics 'Ettore Pancini', University of Naples 'Federico II', Via Cintia, 1-80126 Napoli, Italy
* Correspondence: mariano.palomba@cnr.it (M.P.); ubaldo.coscia@unina.it (U.C.)

Citation: Longo, A.; Di Bartolomeo, A.; Faella, E.; Pelella, A.; Giubileo, F.; Sorrentino, A.; Palomba, M.; Carotenuto, G.; Barucca, G.; Tagliaferro, A.; et al. Temperature Dependence of Electrical Resistance in Graphite Films Deposited on Glass and Low-Density Polyethylene by Spray Technology. *Coatings* **2022**, *12*, 1446. https://doi.org/10.3390/coatings12101446

Academic Editor: Alexander Ulyashin

Received: 17 July 2022
Accepted: 26 September 2022
Published: 30 September 2022

Publisher's Note: MDPI stays neutral with regard to jurisdictional claims in published maps and institutional affiliations.

Abstract: Graphite lacquer was simply sprayed on glass and low-density polyethylene (LDPE) substrates to obtain large area films. Scanning Electron Microscopy (SEM) images, Raman spectra, X Ray Diffraction (XRD) spectra and current-voltage characteristics show that at room temperature, the as-deposited films on different substrates have similar morphological, structural and electrical properties. The morphological characterization reveals that the films are made of overlapped graphite platelets (GP), each composed of nanoplatelets with average sizes of a few tens of nanometers and about forty graphene layers. The thermoresistive properties of the GP films deposited on the different substrates and investigated in the temperature range from 20 to 120 °C show very different behaviors. For glass substrate, the resistance of the film decreases monotonically as a function of temperature by 7%; for LDPE substrate, the film resistance firstly increases more than one order of magnitude in the 20–100 °C range, then suddenly decreases to a temperature between 105 and 115 °C. These trends are related to the thermal expansion properties of the substrates and, for LDPE, also to the phase transitions occurring in the investigated temperature range, as evidenced by differential scanning calorimetry measurements.

Keywords: graphite platelet coatings; low-density polyethylene; thermal expansion coefficient; phase transitions; thermoresistive properties

1. Introduction

The improvement of the electrical conductivity of polymers for their use in flexible electronics [1–6] can be achieved efficiently by coupling them with materials such as carbon black [7], carbon fibers [8–10], graphite [11,12], carbon nanotubes [13], graphene [14,15] or reduced graphene oxide [16,17]. In particular, it is possible to fabricate electrically conductive paths [18], printed radio frequency devices [19], flexible sensors [2,20] and piezoresistive sensors [21] by covering the surfaces of polyethylene terephthalate, low-density polyethylene (LDPE) or poly (methyl methacrylate) with graphite or graphene layers [22–24].

These carbon-based layers can be obtained by using different techniques such as the casting and drying of inks [18], chemical vapor deposition [25], micromechanical techniques utilizing the spreading of graphite nanoplatelets in an alcoholic suspension [22–24] or spraying conductive composites [26,27]. Above all, this last technique is simple, low cost, and suitable for the fabrication of large area conductive films [28] and polymeric film strain gauges suitable for the measurement of large elongations [26].

In this paper, an extensive study is conducted on the properties of graphite platelet (GP) films obtained by spraying a commercial lacquer on different substrates. The structural and morphological characterizations of GP films are performed by a variety of techniques such as differential scanning calorimetry (DSC), X-ray diffraction (XRD), scanning electron microscopy (SEM) and Raman spectroscopy.

The thermoresistive properties of GP deposited on LDPE and glass substrates are investigated by the resistance trends in the 20–120 °C temperature range. In the case of films on LDPE, variations in the thermal expansion coefficient and phase transitions occurring in the polymer substrates are taken into account in the discussion. Understanding the dependence of thermoresistive properties on the thermomechanical ones is crucial in view of large-scale production.

2. Materials and Methods

Using Graphit 33, a commercial lacquer produced by kontakt Chemie (Zelecity, Belgium), spray technology was exploited to deposit large area graphite-based films on glass and LDPE substrates. LDPE substrates of size 15 cm × 15 cm × 90 µm were produced by Sabic (Riyadh, Saudi Arabia, 99.77% by weight, Mn = 280,000 g·mol^{-1}, melt flow index of 7 g/10 min and crystalline fraction (Xc) of 32.8%), while glass substrates (SiO_2) of size 2.5 cm × 7.6 cm × 1 mm were fabricated by Pearl, Inc. (West Hollywood, CA, USA). Figure 1 shows the scheme of the apparatus used for the spray technique.

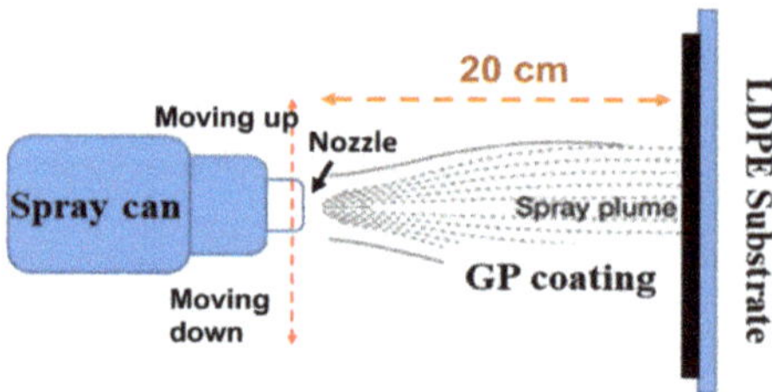

Figure 1. Scheme of the apparatus used for the spray deposition technique.

The spray nozzle was directed to a distance of 20 cm from the plane containing the substrates to produce a uniform deposition by full cone spot and horizontally moved so that a graphite film could completely cover the surface of the substrates. Four runs of depositions were performed, and for each run, a pair of graphite films was obtained at room temperature (RT) under the same deposition conditions on glass and LDPE substrates.

Before lacquer spraying, no cleaning treatment was performed on the LDPE substrates, while glass substrates were cleaned by acetone. After spraying, the coated substrates were kept in air at room temperature to dry for 4 h.

The graphite content in the Graphite 33 lacquer measured by TGA analysis corresponded to 82% by weight, as indicated in our previous article (see Figure 2 in Ref. [27]) and the as deposited amount of lacquer equal to 7.20 mg/cm^2 became 0.983 mg/cm^2 (0.806 mg/cm^2 of graphite) after drying in air for 4 h.

A Raman spectrometer, InviaH-Renishaw (New Mils, Wotton-under Edge, Glowcester Shire, GL128JR, UK), was employed to acquire the Raman spectra of the GP films on both glass and LDPE substrates. The analysis was carried out using a green argon laser with 514.5 nm wavelength and about 2 µm diameter beam spot. The microscope was operated at 50× magnification and 10 s exposure time. Using ~5 mW laser power (corresponding to 5% of the available laser power), 100 to 3500 cm^{-1} scans were performed. The quantum efficiency of the detector was corrected by a dedicated MATLAB software

(Matlab 9.7.0. 1190202 R2019b, Mathworks Inc., Natick, MA, USA), which was also used for the baseline subtraction and the data processing.

The analysis of the samples surface was performed by scanning electron microscopy by means of a Zeiss Supra 40 field-emission SEM (Carl Zeiss Microscopy GmbH, Jena, Germany). The samples were accurately attached on stubs and scrutinized without any preparation. As graphite-based materials are conductive, their surface was electrically grounded.

The mean thickness value of the investigated films, estimated by transmission electron microscopy observations, as reported elsewhere [23], was about 2.5 µm.

Sample structural properties were also studied by X-ray diffraction (XRD) measurements performed by a Panalytical, X'PERT PRO diffractometer (Malvern Panalytical, Cambridge, UK) with $\lambda = 0.154$ nm CuKα radiation, 5°–60° 2θ range, 0.0130° scan step and 18.9 s full scan time.

Phase transitions of the LDPE substrate were investigated by differential scanning calorimetry (DSC) using a Q1000-TA instrument (TA instrument, New Castle, PA, USA). The apparatus was calibrated using indium and zinc as reference and non-isothermal cooling-heating-cooling tests were carried out at different scan rates in the 20–180 °C temperature range.

A fully automated GNOMIX high-pressure mercury dilatometer was used for Pressure-Volume-Temperature (PVT) experiments to explore the thermal expansion properties of the LDPE substrate [29]. The instrument was able to detect changes in specific volume with an absolute accuracy of 0.004 cm$^3\cdot$g^{-1}. The polymer density variations were measured from room temperature to 137 °C with the scanning rate of 3 °C/min. The thermal linear strain, ε, of the LDPE was determined in heating mode from the changes in volume obtained during isobaric PVT runs as:

$$\varepsilon = \upsilon \times (V - V_0)/V_0 \tag{1}$$

where υ is the Poisson modulus equal to 0.46 for LDPE, V_0 is the initial volume of the sample at the temperature $T_0 = 20$ °C and V is the volume of the sample at the temperature, T. Resistance measurements of the deposited graphite film versus temperature, in the 20–120 °C range, were performed under 2 mbar vacuum in a coplanar configuration, after spreading the sample surfaces by 1 cm long and 1 mm spaced silver-paint contacts.

A source-measurement unit (SMU) Keithley 4200-SCS (Tektronix, Inc., Beaverton, OR, USA), electrically connected to the four micromanipulators of a Janis Research ST-500 probe station (Janis Research, Woburn, MA, USA), was used to measure the resistances of the samples as mean values from the I-V characteristics taken at fixed temperatures.

3. Results

Films deposited on different substrates by spraying Graphit 33 lacquer under the same preparation conditions were investigated for their morphological, structural and thermoresistive properties. After spraying the lacquer, a good adhesion of the graphite film takes place on LDPE substrate because graphite platelets interact with the surface of non-polar materials, such as polyethylene. On the other hand, the polymeric binder present in the lacquer allows the adhesion of the graphite film to the glass substrate.

The Raman spectra of two representative samples prepared on glass and LDPE substrates are plotted in Figure 2a,b, respectively. No substrate signals are detected as the films' thicknesses are higher than the penetration length of the laser radiation used in the Raman apparatus, i.e., a few hundred nanometers. Hence, the Raman spectra are due to the graphite platelets embedded in the lacquer. The features in the 1000–1800 cm^{-1} region are quite similar in the two spectra: a narrow D peak, a sharp and narrow G band with a shoulder (D*) due to the presence of defects [30]. The peaks are sharp, indicating that the graphite platelets are of good crystalline quality. The peak intensity ratio, I_d/I_g, averaged over the sampled points, are 0.55 and 0.80 for the films on glass and LDPE, respectively. It is worthwhile to note that the D peak is generated by the platelet edges [31], and indeed in the case of a platelet slightly larger than the beam and perpendicular to it, no edges will be excited and no D peak will be detected [31]. On the other end, for bended platelet not perpendicular to the beam, some edges will be illuminated producing a D peak. The results indicate that more edges fall into the area swept by the laser beam in the LDPE case. This is consistent with a slightly more disorder in orientation of the platelets on the LDPE substrate.

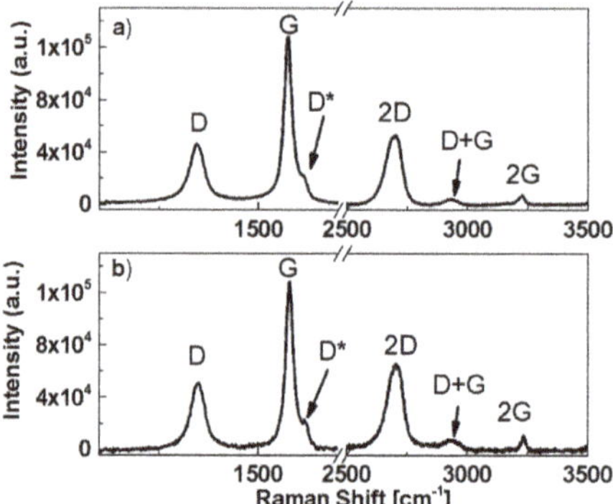

Figure 2. Raman spectra of films deposited by Graphit 33 lacquer on glass (**a**) and LDPE (**b**).

Three further peaks can be detected at higher wavenumbers between 2500 and 3500 cm^{-1} in both spectra. The intensity of the left shouldered 2D peak indicates a multilayered structure, while the D + G peak and the 2G peak have the typical width and intensity of graphite material.

The surface morphology of the samples was investigated by SEM. The micrographs of two films deposited on glass and LDPE substrates are shown in Figure 3a,b, respectively. SEM micrographs reveal that the films on both substrates are composed of aggregated graphite platelets (GP) having a tendency to align parallel to the LDPE/film interface but arranged in a rather disordered way and partially overlapping. GP are irregular in shape, with a broad distribution of lateral dimensions ranging from a few hundreds of nanometers to a few micrometers. Although the platelets tend to align parallel to the interfacial plane, it can be observed that platelets tilted with respect to the substrate plane (as in the insets of Figure 3) make the surface of the films rough and porous. SEM analysis reveals that films deposited on the different substrates exhibit a similar surface structure.

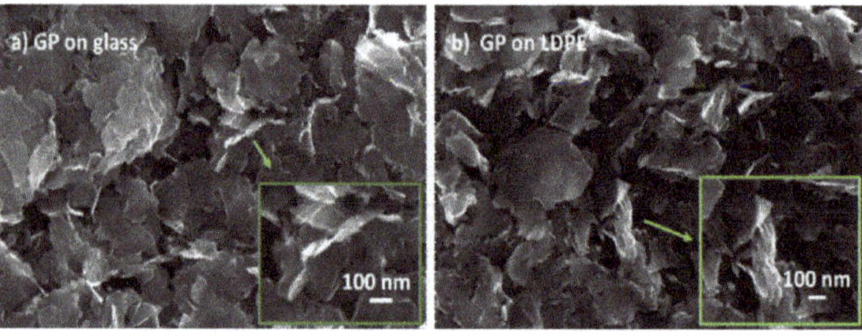

Figure 3. SEM micrograph of GP films deposited on glass (**a**) and LDPE (**b**). The insets show a magnification of platelets tilted with respect to the substrate planar surface.

Further information on the structure of the GP films is obtained through XRD measurements. The XRD spectrum of the film deposited on glass is shown in Figure 4a. It includes only one peak centered at $2\theta = 26.46$, attributed to the (002) crystallographic planes of graphite and the typical amorphous halo due to the glass substrate [International Center for diffraction Data file: 00-008-0415].

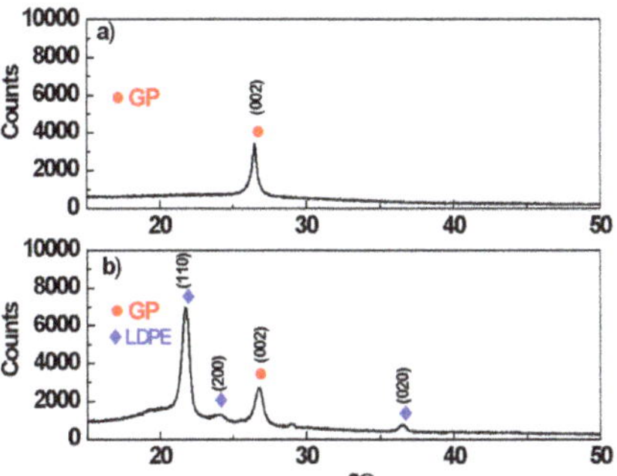

Figure 4. XRD diffractograms of graphite platelet (GP) films deposited on glass (**a**) and LDPE (**b**).

On the other hand, the XRD diffractogram of the film deposited on LDPE and displayed in Figure 4b includes both LDPE and graphite diffraction peaks. Three peaks assigned to the crystalline phase of LDPE [32], superimposed on a large halo generated by the amorphous polyethylene phase [33,34], can be identified: a dominant sharp peak at $2\theta = 21.79°$, a weak broad peak at $2\theta = 24.05°$ and one more weak peak at $2\theta = 36.57°$ corresponding, respectively, to (110), (200) and (020) crystallographic planes. In addition, a peak of low intensity at $2\theta = 26.74°$ due to the graphite (002) plane [ICDD: 75-1621] is also detected.

The average thickness of the graphite crystals, $L_{(hkl)}$, perpendicular to the lattice plane identified by Miller's indices (hkl), is evaluated using Scherrer's equation:

$$L_{(hkl)} = K \times \lambda / (FWHM \times \cos(\theta)) \tag{2}$$

where θ is the half of the corresponding scattering angle of peak (hkl), FWHM in radians is the full-width at half-maximum and stems from the Gaussian fit of the (hkl) peak, K assumed to be 0.89 is a constant dependent on the crystallite shape [35,36], and λ is the wavelength of the X-ray radiation (Cu-$K_{\alpha1}$ = 0.15481 nm). The Equation (2) applied to the (002) peak of graphite yields $L_{(hkl)} = 15.2 \pm 0.5$ nm and $L_{(hkl)} = 12.6 \pm 0.4$ nm for films deposited on glass and LDPE, respectively. Moreover, the inter-plane distance is obtained by Bragg's law for a family of lattice planes (hkl) [37,38]:

$$d_{(hkl)} = n\lambda / \sin(\theta) \tag{3}$$

where $n = 1$ is the diffraction peak order. The calculated values are $d_{(002)} = 0.337 \pm 0.002$ nm and $d_{(002)} = 0.333 \pm 0.003$ nm for the films on glass and LDPE, respectively, in agreement with the separation distance between two graphene layers (d = 0.335 nm) inside the graphite phase [39].

Finally, the average number of the graphene layers (N) per crystalline domain is determined from the following equation:

$$N = (L_{(002)} / d_{(002)}) + 1 \tag{4}$$

The calculated N values are 46 ± 1 and 38 ± 1 for the samples deposited on glass and LDPE, respectively. Thus, it can be stated that the deposited films are made of an aggregate of overlapping graphite platelets, each formed by nanoplatelets with average sizes of a few tens of nanometers and about forty graphene layers.

Regarding the thermoresistive properties of GP films, the influence of the substrate thermomechanical parameters has to be considered. Since the properties of glass are well known, only the temperature ranges, in which phase transitions of LDPE occur and the coefficient of linear thermal expansion are determined in this work.

Figure 5 shows the conventional DSC results from the heating of a representative LDPE substrate in the 20–180 °C range at the scanning rate of 10 °C/min.

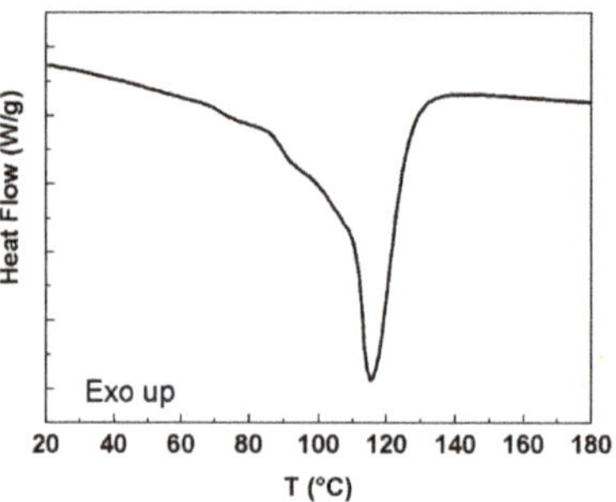

Figure 5. DSC heating curves of a LDPE sample in the 20–180 °C temperature, T, range at the rate of 10 °C/min.

LDPE is a ductile and flexible material at room temperature. It is solid and stable in the temperature range from 20 to 85 °C, while it presents a complex phenomenon resulting from a broad melting transition and reorganization of the crystalline phase if the temperature is greater than 85 °C and less than or equal to 120 °C, as displayed in Figure 4 by the enthalpic curve having a minimum value at about 114 °C. At temperatures above 120 °C, the LDPE first becomes "sticky" (120–140 °C) and then a liquid melt (140–180 °C). As with any other polymeric material, this temperature range is characterized by a partial rearrangement of the molecular chains with the formation of a high viscous liquid with some elastic characteristics (viscoelastic behavior). During this phase transition, the material becomes sticky because it has some typical characteristics of liquids and therefore "wets" the surfaces with which it is in contact; but then, due to its elasticity, it will resist separation when stressed [40].

The thermal expansion properties of this polymer are studied by means of the thermal linear strain, ε, of a typical LDPE substrate plotted, in Figure 6, as a function of the temperature, T, in the 20–137 °C range. Here, ε is defined as:

$$\varepsilon = (L - L_0)/L_0 \tag{5}$$

where L_0 is the initial length of the sample at the temperature $T_0 = 20$ °C and L is the length of the sample at the temperature, T.

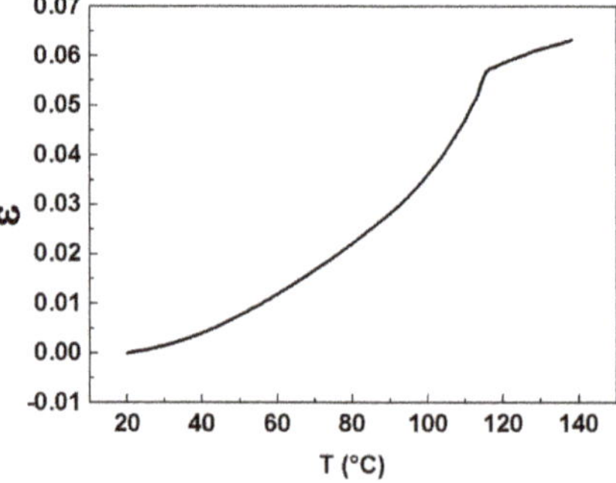

Figure 6. Linear strain ε vs temperature, T, of an LDPE substrate.

Clearly, the LDPE thermal expansion is not proportional to the change in temperature, and therefore the linear coefficient of the thermal expansion, CTE, at a given temperature T is calculated as:

$$CTE = \frac{1}{L}\frac{dL}{dT} \tag{6}$$

where L is the length of the sample at the temperature T and dL/dT is the rate of change of the linear dimension per unit change in temperature.

The CTE versus T plot is reported in Figure 7.

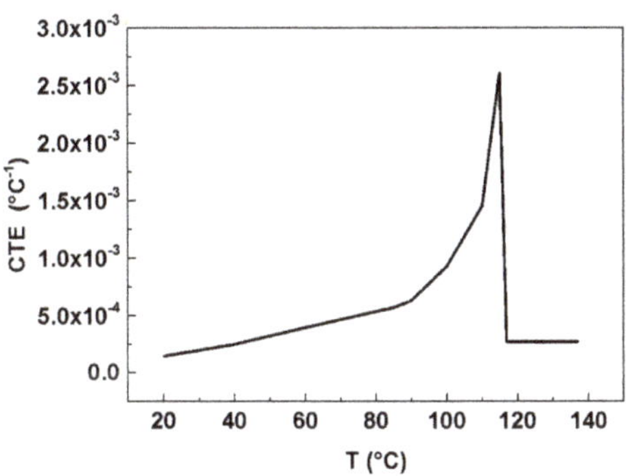

Figure 7. The coefficient of linear thermal expansion, CTE, vs. temperature, T, of an LDPE substrate.

The CTE at $T_0 = 20\,^\circ$C is $1.5 \times 10^{-4}\,^\circ$C^{-1} in agreement with data in the literature [41]. Furthermore, by correlating the CTE trend with the DSC results one obtains that in the 20–85 $^\circ$C range, where the LDPE phase is solid and stable, the CTE increases between 1.5×10^{-4} and $5.7 \times 10^{-4}\,^\circ$C^{-1}, while if the temperature is greater than 85 $^\circ$C, where a broad melting transition takes place, CTE increases faster, up to $2.6 \times 10^{-3}\,^\circ$C^{-1} near 114 $^\circ$C^{-1}. Finally, the CTE suddenly drops to a constant value of $2.7 \times 10^{-4}\,^\circ$C^{-1} from 117 to 137 $^\circ$C.

In the investigated temperature range, this polymer is an excellent insulator with good dielectric properties and a high-volume resistance, but its surface can be easily made electrically conductive by depositing a GP film by spraying Graphit 33 lacquer, as shown in Figure 8, which displays the current-voltage (I-V) characteristics of a LDPE sample sprayed with Graphit 33 lacquer. The measurements are performed in vacuum in two-probe configuration at the temperature of 20 $^\circ$C. Figure 8 also shows the I-V characteristics of a GP film produced under the same deposition conditions on glass substrate.

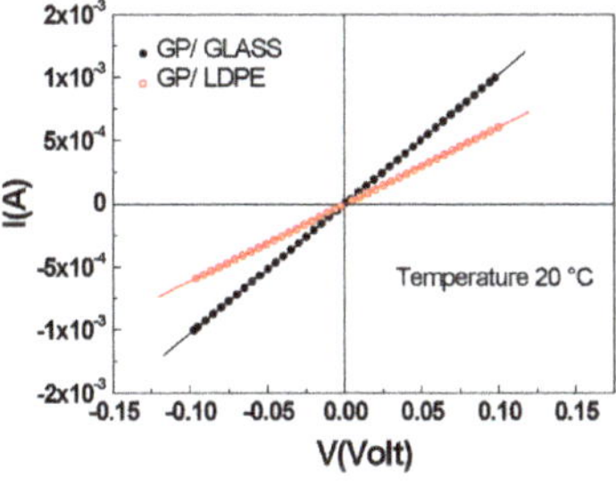

Figure 8. I-V characteristics of the GP films deposited on glass (black circle symbols) and LDPE (red circle symbols) substrates. The straight lines fit the experimental data.

The I-V characteristics of both samples are linear, indicating ohmic contacts. The resistance values, R_0, resulting from the best fit of the measured I-V curve with a correlation coefficient r = 1, are (97.49 $\pm$ 0.02) Ω for GP deposited on glass and (164.00 $\pm$ 0.03) Ω for GP on LDPE.

The thermoresistive properties of the GP samples are explored by measuring the resistance R while raising the temperature from 20 to 120 °C. In Figure 9, the R/R_0 ratios are displayed as a function of the temperature, T.

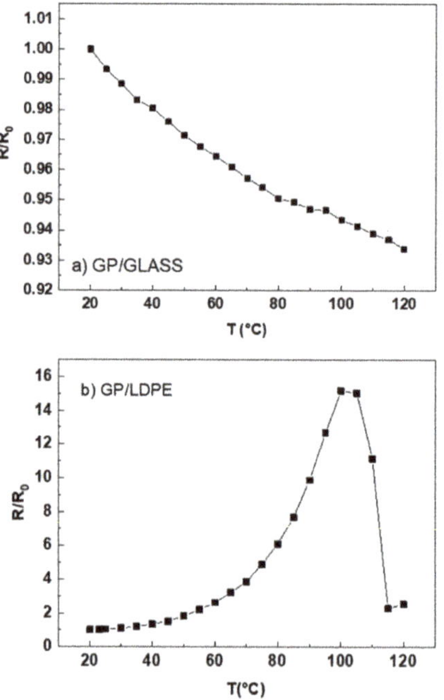

Figure 9. Ratio of the resistance R to the initial resistance R_0 at 20 °C (R/R_0) vs. temperature, T, for GP films deposited on glass (**a**) and LDPE (**b**) substrates. Black lines are guides.

Different thermoresistive behaviors can be noted. An increasing temperature corresponds to a decreasing resistance of GP film on glass (Figure 9a). Conversely, for GP on LDPE, the resistance rises strongly, more than one order of magnitude, in the range from 20 °C to about 100 °C, and after reaching the maximum value, it suddenly decreases between 105 and 115 °C (Figure 9b). These trends can be ascribed to the different thermomechanical properties of the substrates. Indeed, glass has a linear thermal expansion coefficient, CTE = $6 - 9 \times 10^{-6}$ °C^{-1} [42] close to that of graphite ($4 - 8 \times 10^{-6}$ °C^{-1} [42]); therefore, the thermal expansion properties of the substrate do not affect those of the film, the resistance of which decreases with the rising to T, as in graphite [43].

On the other hand, as determined by the thermal expansion experiments, the CTE of the LDPE substrate at $T_0 = 20$ °C is 1.5×10^{-6} °C^{-1}, thus more than one order of magnitude greater than in graphite, and it increases further as a function of temperature up to 114 °C, and then CTE assumes a constant value of 2.7×10^{-6} °C^{-1} between 117 and 137 °C.

Interestingly, from 20 to 100 °C, the trend of resistance ratio is correlated with the increasing mismatch between LDPE substrate and GP film because in the 20–85 °C temperature range R/R_0 increases slowly as the CTE of LDPE, while if the temperature is greater than 85 °C and less than or equal to 100 °C, the slope of the R/R_0 curve increases sharply as in the case of CTE-T graph shown in Figure 7.

Consequently, the resistance increase observed in the 20–100 °C range, as visible in Figure 9b, can be attributed to the thermal expansion coefficient mismatch between the LDPE and GP films.

Indeed, the higher thermal expansion of the polymeric substrate can induce strains in the deposited film that decrease the contact area among the platelets and cause the occurrence of micro/nanofractures. These occurrences reduce the number of conductive paths and the film resistance increases.

The decrease in resistance between 105 and 115° C, despite the increase in the polymer substrate's thermal expansion, can be due to the fact that in this range a significant fraction of the substrate has melted, resulting in a sharp decrease in the mechanical modulus of the LDPE which favors the sliding of the graphite platelets and their redistribution with the formation of new conductive paths on the sample surfaces and the consequent decrease in the film resistance.

The electrical conductivity of the LDPE filled with carbon fiber composites carbon or black showed a similar temperature dependence [44]. The investigated GP films on LDPE exhibit significant thermoresistive response and have potential applications for flexible electronics, including temperature sensors and self-switching components.

4. Conclusions

Large area graphite platelets films were deposited on glass and LDPE substrates by a spray technology using a commercial graphite lacquer. Analyses of Raman spectra, SEM images and XRD diffractograms show that the films are made of overlapped graphite platelets each composed of nanoplatelets with average size of a few tens of nanometers and forty graphene layers. The enthalpic curve obtained by DSC measurements of LDPE indicates that this polymer is solid and stable in the 20–85 °C range, while it undergoes a broad melting transition in the temperature range between 85 and 120 °C. In particular, the endothermic peak shows a minimum at about 114 °C. It has been found that the thermal expansion properties of LDPE are strongly dependent on the temperature. Indeed, in the 20–85 ° C range, the CTE increases between 1.5×10^{-4} and $5.7 \times 10^{-4} \, °C^{-1}$, while if the temperature is greater than 85 °C, CTE increases faster, up to $2.6 \times 10^{-3} \, °C^{-1}$ near 114 °C, and then it drops to a constant value of $2.7 \times 10^{-4} \, °C^{-1}$ from 117 to 137 °C.

The thermoresistive properties of the GP film are affected by the mismatch of CTE between the film and the substrate. Indeed, the CTE of the glass substrate is close to that of the graphite film, and the thermal expansion properties of the substrate do not influence those of the film, the resistance of which decreases with the increasing of the temperature, as in graphite. Conversely, for GP deposited on LDPE substrate, the increasing CTE mismatch between the film and the substrate, in the 20–100 °C range, induces strains in the film that decreases the contact area among the platelets and increases the nano/micro-fractures, thus resulting in the resistance increase of the films. Finally, between 105 and 115 °C, the melting of the polymer substrate favors the sliding of the graphite platelets with the formation of new conductive paths that decrease the film resistance.

Author Contributions: Conceptualization, U.C., G.C., A.S., G.B. and A.D.B.; software, A.L., A.S., G.B. and M.P.; validation, U.C., A.L., M.P., A.S., G.B. and A.T.; formal analysis, A.L., M.P., U.C., A.S. and G.B.; investigation, G.B., E.F., A.P., F.G., A.L., A.S. and M.P.; data curation, G.C., A.L. and M.P.; writing—original draft preparation, G.C., M.P., A.L., A.D.B., G.B., A.S., E.F., A.P., F.G., A.T. and U.C.; writing—review and editing A.L., A.D.B., E.F., A.P., F.G., A.S., M.P., G.C., G.B., A.T. and U.C.; supervision, U.C., A.S., G.C. and A.D.B. All authors have read and agreed to the published version of the manuscript.

Funding: This research received no external funding.

Institutional Review Board Statement: Not applicable.

Informed Consent Statement: Not applicable.

Data Availability Statement: Not applicable.

Acknowledgments: We wish to acknowledge the contribution of the late Massimo Rovere to this paper, unfortunately one of his last contributions to science.

Conflicts of Interest: The authors declare no conflict of interest.

References

1. Corzo, D.; Tostado-Blázquez, G.; Baran, D. Flexible electronics: Status, challenges and opportunities. *Front. Electron.* **2020**, *1*, 594003. [CrossRef]
2. Costa, J.C.; Spina, F.; Lugoda, P.; Garcia-Garcia, L.; Roggen, D.; Münzenrieder, N. Flexible sensors: From materials to applications. *Technologies* **2019**, *7*, 35. [CrossRef]
3. Seo, C.U.; Yoon, Y.; Kim, D.H.; Choi, S.Y.; Park, W.K. Fabrication of polyaniline-carbon nano composite for application in sensitive flexible acid sensor. *J. Ind. Eng. Chem.* **2018**, *64*, 97–101. [CrossRef]
4. Wang, P.; Hu, M.; Wang, H.; Chen, Z.; Feng, Y.; Wang, J.; Ling, W.; Huang, Y. The evolution of flexible electronics: From nature, beyond nature, and to nature. *Adv. Sci.* **2020**, *7*, 2001116. [CrossRef]
5. Hussain, M.M.; El-Atab, N. *Handbook of Flexible and Stretchable Electronics*, 1st ed.; CRC Press: Boca Raton, FL, USA, 2019.
6. Garidiner, F.; Carter, E. *Polymer Electronics—A Flexible Technology*; iSmithers Rapra Publishing: Akron, OH, USA, 2010.
7. Shintake, J.; Piskarev, E.; Jeong, S.H.; Floreano, D. Ultrastretchable strain sensors using carbon black-filled elastomer composites and comparison of capacitive versus resistive sensors. *Adv. Mater. Technol.* **2017**, *3*, 1700284. [CrossRef]
8. Wang, S.; Kowalik, D.P.; Chung, D.D.L. Self-sensing attained in carbon-fiber–polymer-matrix structural composites by using the interlaminar interface as a sensor. *Smart Mater. Struct.* **2004**, *13*, 570–592. [CrossRef]
9. Dong, Q.; Guo, Y.; Sun, X.; Jia, Y. Coupled electrical-thermal-pyrolytic analysis of carbon fiber/epoxy composites subjected to lightning strike. *Polymer* **2015**, *56*, 385–394. [CrossRef]
10. Sibinski, M.; Jakubowska, M.; Sloma, M. Flexible temperature sensors on fibers. *Sensors* **2010**, *10*, 7934–7946. [CrossRef]
11. Tian, X.; Itkis, M.E.; Bekyarova, E.B.; Haddon, R.C. Anisotropic thermal and electrical properties of thin thermal interface layers of graphite nanoplatelet-based composites. *Sci. Rep.* **2013**, *3*, 1710. [CrossRef]
12. Nasim, M.N.E.A.; Chun, D.-M. Substrate-dependent deposition behavior of graphite particles dry-sprayed at room temperature using a nano-particle deposition system. *Surf. Coat. Technol.* **2017**, *309*, 172–178. [CrossRef]
13. Shi, J.; Li, X.; Cheng, H.; Liu, Z.; Zhao, L.; Yang, T.; Dai, Z.; Cheng, Z.; Shi, E.; Yang, L.; et al. Graphene reinforced carbon nanotube networks for wearable strain sensors. *Adv. Funct. Mater.* **2016**, *26*, 2078–2084. [CrossRef]
14. Novoselov, K.S.; Geim, A.K.; Morozov, S.V.; Jiang, D.; Zhang, Y.; Dubonos, S.V.; Grigorieva, I.V.; Firsov, A.A. Electric field effect in atomically thin carbon films. *Science* **2004**, *306*, 666–669. [CrossRef] [PubMed]
15. Allen, M.J.; Tung, V.C.; Kaner, R.B. Honeycomb carbon: A review of graphene. *Chem. Rev.* **2010**, *110*, 132–145. [CrossRef] [PubMed]
16. Longo, A.; Verucchi, R.; Aversa, L.; Tatti, R.; Ambrosio, A.; Orabona, E.; Coscia, U.; Carotenuto, G.; Maddalena, P. Graphene oxide prepared by graphene nano-platelets and reduced by laser treatment. *Nanotechnology* **2017**, *28*, 224002. [CrossRef] [PubMed]
17. Soni, M.; Kumar, P.; Pandey, J.; Sharma, S.K.; Soni, A. Scalable and site specific functionalization of reduced graphene oxide for circuit elements and flexible electronics. *Carbon* **2018**, *128*, 172–178. [CrossRef]
18. Liu, H.; Gao, J.; Huang, W.; Dai, K.; Zheng, G.; Liu, C.; Shen, C.; Yan, X.; Guo, J.; Guo, Z. Electrically conductive strain sensing polyurethane nanocomposites with synergistic carbon nanotubes and graphene bifillers. *Nanoscale* **2016**, *8*, 12977–12989. [CrossRef]
19. Huang, X.; Leng, T.; Zhang, X.; Chen, J.C.; Chang, K.H.; Geim, A.K.; Novoselov, K.S.; Hu, Z. Binder-free highly conductive graphene laminate for low cost printed radio frequency applications. *Appl. Phys. Lett.* **2015**, *106*, 203105. [CrossRef]
20. Liu, G.; Tan, Q.; Kou, H.; Zhang, L.; Wang, J.; Lv, W.; Dong, H.; Xiong, J. A flexible temperature sensor based on reduced graphene oxide for robot skin used in internet of things. *Sensors* **2018**, *18*, 1400. [CrossRef]
21. Bonavolontà, C.; Camerlingo, C.; Carotenuto, G.; De Nicola, S.; Longo, A.; Meola, C.; Boccardi, S.; Palomba, M.; Pepe, G.P.; Valentino, M. Characterization of piezoresistive properties of graphene-supported polymer coating for strain sensor applications. *Sens. Actuators Phys.* **2016**, *252*, 26–34. [CrossRef]
22. Palomba, M.; Longo, A.; Carotenuto, G.; Coscia, U.; Ambrosone, G.; Rusciano, G.; Nenna, G.; Barucca, G.; Longobardo, L. Optical and electrical characterizations of graphene nanoplatelet coatings on low density polyethylene. *J. Vac. Sci. Technol. B Nanotechnol. Microelectron. Mater. Process. Meas. Phenom.* **2018**, *36*, 01A104. [CrossRef]
23. Coscia, U.; Palomba, M.; Ambrosone, G.; Barucca, G.; Cabibbo, M.; Mengucci, P.; de Asmundis, R.; Carotenuto, G. A new micromechanical approach for the preparation of graphene nanoplatelets deposited on polyethylene. *Nanotechnology* **2017**, *28*, 194001. [CrossRef] [PubMed]
24. Coscia, U.; Longo, A.; Palomba, M.; Sorrentino, A.; Barucca, G.; Di Bartolomeo, A.; Urban, F.; Ambrosone, G.; Carotenuto, G. Influence of the thermomechanical characteristics of low-density polyethylene substrates on the thermoresistive properties of graphite nanoplatelet coatings. *Coatings* **2021**, *11*, 332. [CrossRef]
25. De Castro, R.K.; Araujo, J.R.; Valaski, R.; Costa, L.O.O.; Archanjo, B.S.; Fragneaud, B.; Cremona, M.; Achete, C.A. New transfer method of CVD-grown graphene using a flexible, transparent and conductive polyaniline-rubber thin film for organic electronic applications. *Chem. Eng. J.* **2015**, *273*, 509–518. [CrossRef]
26. Kondratov, A.P.; Zueva, A.M.; Varakin, R.S.; Taranec, I.P.; Savenkova, I.A. Polymer film strain gauges for measuring large elongations. *IOP Conf. Ser. Mater. Sci. Eng.* **2018**, *312*, 012013. [CrossRef]
27. Palomba, M.; Carotenuto, G.; Longo, A.; Sorrentino, A.; Di Bartolomeo, A.; Iemmo, L.; Urban, F.; Giubileo, F.; Barucca, G.; Rovere, M.; et al. Thermoresistive properties of graphite platelet films supported by different substrates. *Materials* **2019**, *12*, 3638. [CrossRef]

28. Di Bartolomeo, A.; Iemmo, L.; Urban, F.; Palomba, M.; Carotenuto, G.; Longo, A.; Sorrentino, A.; Giubileo, F.; Barucca, G.; Rovere, M.; et al. Graphite platelet films deposited by spray technique on low density polyethylene substrates. *Mater. Today Proc.* **2020**, *20*, 87–90. [CrossRef]
29. Walsh, D.; Zoller, P. *Standard Pressure Volume Temperature Data for Polymers*; CRC Press: Boca Raton, FL, USA, 1995.
30. Ferrari, A.C.; Robertson, J. Interpretation of Raman spectra of disordered and amorphous carbon. *Phys. Rev. B* **2000**, *61*, 14095. [CrossRef]
31. Pimenta, M.A.; Dresselhaus, G.; Dresselhaus, M.S.; Canc, L.G.; Jorioa, A.; Saitoe, R. Studying disorder in graphite-based systems by Raman spectroscopy. *Phys. Chem. Chem. Phys.* **2007**, *9*, 1276–1291. [CrossRef]
32. Alexander, L.E. *X-ray Diffraction Methods in Polymer Science*; Krieger Publishing: Huntington, NY, USA, 1979.
33. Singh, B.P.; Saini, P.P.; Gupta, T.; Garg, P.; Kumar, G.; Pande, I.; Pande, S.; Seth, R.K.; Dhawan, S.K.; Mathur, R.B. Designing of multiwalled carbon nanotubes reinforced low density polyethylene nanocomposites for suppression of electromagnetic radiation. *J. Nanopart. Res.* **2011**, *13*, 7065–7074. [CrossRef]
34. Akinci, A. Mechanical and morphological properties of basalt filled polymer matrix composites. *Arch. Mater. Sci. Eng.* **2009**, *35*, 29–32.
35. Sharma, R.; Chadha, N.; Saini, P. Determination of defect density, crystallite size and number of graphene layers in graphene analogues using X-ray diffraction and Raman spectroscopy. *Indian J. Pure Appl. Phys.* **2017**, *55*, 625–629.
36. Buerger, M.J. *X-ray Crystallography*; Wiley: Hoboken, NJ, USA, 1942.
37. Cullity, B.D. *Elements of X-ray Diffraction*; Addison-Wesley: Boston, MA, USA, 1956.
38. Bacon, G.E. Unit-cell dimensions of graphite. *Acta Cryst.* **1950**, *3*, 137. [CrossRef]
39. Bacon, G.E. The rhombohedral modification of graphite. *Acta Cryst.* **1950**, *3*, 320. [CrossRef]
40. Kalika, D.S.; Morton, M.D. Wall slip and extrudate distortion in linear low-density polyethylene. *J. Rheol.* **1987**, *31*, 815–834. [CrossRef]
41. Kissin, Y.V. *Polyethylene: End-Use Properties and Their Physical Meaning*; Hanser Pub Inc.: Liberty Township, OH, USA, 2013.
42. Thermal Expansion—Linear Expansion Coefficients. Available online: https://www.engineeringtoolbox.com/linear-expansion-coefficients-d_95.html (accessed on 16 July 2022).
43. Iwashita, H.; Imagawa, H.; Nishiumi, W. Variation of temperature dependence of electrical resistivity with crystal structure of artificial products. *Carbon* **2013**, *61*, 602–608. [CrossRef]
44. Di, W.; Zhang, G.; Xu, J.; Peng, Y.; Wang, X.; Xie, Z. Positive-temperature-coefficient/negative-temperature-coefficient effect of low-density polyethylene filled with a mixture of carbon black and carbon fiber. *J. Polym. Sci. Part B Polym. Phys.* **2003**, *41*, 3094–3101. [CrossRef]

Communication

Structural and Chemical Peculiarities of Nitrogen-Doped Graphene Grown Using Direct Microwave Plasma-Enhanced Chemical Vapor Deposition

Šarūnas Meškinis , Rimantas Gudaitis, Mindaugas Andrulevičius and Algirdas Lazauskas *

Institute of Materials Science, Kaunas University of Technology, K. Baršausko 59, LT-51423 Kaunas, Lithuania; sarunas.meskinis@ktu.lt (Š.M.); rimantas.gudaitis@ktu.lt (R.G.); mindaugas.andrulevicius@ktu.lt (M.A.)
* Correspondence: algirdas.lazauskas@ktu.edu; Tel.: +370-671-73375

Abstract: Chemical vapor deposition (CVD) is an attractive technique which allows graphene with simultaneous heteroatom doping to be synthesized. In most cases, graphene is grown on a catalyst, followed by the subsequent transfer process. The latter is responsible for the degradation of the carrier mobility and conductivity of graphene due to the presence of the absorbants and transfer-related defects. Here, we report the catalyst-less and transfer-less synthesis of graphene with simultaneous nitrogen doping in a single step at a reduced temperature (700 °C) via the use of direct microwave plasma-enhanced CVD. By varying nitrogen flow rate, we explored the resultant structural and chemical properties of nitrogen-doped graphene. Atomic force microscopy revealed a more distorted growth process of graphene structure with the introduction of nitrogen gas—the root mean square roughness increased from 0.49 ± 0.2 nm to 2.32 ± 0.2 nm. Raman spectroscopy indicated that nitrogen-doped, multilayer graphene structures were produced using this method. X-ray photoelectron spectroscopy showed the incorporation of pure pyridinic N dopants into the graphene structure with a nitrogen concentration up to 2.08 at.%.

Keywords: microwave; plasma-enhanced; CVD; nitrogen-doped; graphene; catalyst-less; transfer-less; synthesis

Citation: Meškinis, Š.; Gudaitis, R.; Andrulevičius, M.; Lazauskas, A. Structural and Chemical Peculiarities of Nitrogen-Doped Graphene Grown Using Direct Microwave Plasma-Enhanced Chemical Vapor Deposition. *Coatings* **2022**, *12*, 572. https://doi.org/10.3390/coatings12050572

Academic Editor: Luca Valentini

Received: 29 March 2022
Accepted: 20 April 2022
Published: 22 April 2022

Publisher's Note: MDPI stays neutral with regard to jurisdictional claims in published maps and institutional affiliations.

1. Introduction

Graphene belongs to a class of two-dimensional (2D) materials, which are widely known for their unique structures and outstanding physical, chemical, and mechanical properties [1–7]. It was demonstrated in many studies and reviews that the properties of 2D materials can be drastically altered, enhanced, or tuned via molecular and atomic doping [8–11]. For instance, the carrier concentration and type of carrier can be easily changed through the substitution of dopant atoms on the sulfur site in titanium trisulfide without having any impact on the band extrema [12]. After doping carbon nanotubes with N or B atoms, they become n-type or p-type atoms, respectively [13]. Doped graphene offers unique properties such as ferromagnetism [14], superconductivity [15], etc. Specifically, graphene heteroatom doping methods can be categorized [16] into post-treatment approaches, e.g., wet chemical methods [17], thermal annealing, arc-discharge [18], plasma treatment [19], hydrothermal treatment [20], and gamma irradiation [21], and in situ approaches, e.g., chemical vapor deposition (CVD) [22], bottom-up synthesis [23], and ball milling [24]. Methods which fall into the latter category are more favorable, as graphene synthesis can be achieved with simultaneous heteroatom doping [16]. Depending on the choice of heteroatoms for doping, e.g., B [25], N [26], P [27], S [28], F [29], Cl [30], Br [31], I [32], etc., new or improved properties of graphene materials may arise and could be useful for a number of applications, including supercapacitors [33], fuel cells [34], lithium ion batteries [35], solar cells [36], and sensors [37].

The CVD technique can be considered as the most effective approach used for graphene doping that does not affect its crystalline nature [38]. Generally, graphene is grown via CVD on a catalyst substrate, e.g., Ni or Cu [39,40], and afterwards, a complicated transfer process is required for the wet chemical removal of the metallic catalyst and the transfer of the graphene onto the required surface based on its intended functionality and application [41]. Importantly, the transfer process is known to be responsible for the generation of defects and the degradation of the carrier mobility and conductivity of graphene, since chemical alterations (i.e., during the wet chemical removal of catalyst) and mechanical damage (i.e., during the transfer process) are almost unavoidable during this process [42]. Additionally, the graphene transfer process is known to introduce unwanted metallic impurities which alter the electrochemical properties of graphene [43]. The demand for monocrystalline Si(1 0 0) continues to rise, as it is a major substrate used in semiconductor device fabrication and optoelectronics. The catalyst-less and transfer-less synthesis of graphene on monocrystalline Si(1 0 0) is meaningful in this context.

Herein, we focused our efforts on demonstrating that the catalyst-less and transfer-less synthesis of graphene can be achieved with simultaneous nitrogen doping in a single step via the use of a direct microwave plasma-enhanced CVD. Importantly, nitrogen-doped graphene synthesis was achieved at a considerably low temperature of 700 °C. By varying nitrogen flow rate, we explored the resultant structural and chemical properties of nitrogen-doped graphene through atomic force microscopy (AFM), Raman spectroscopy, and X-ray photoelectron spectroscopy (XPS).

2. Materials and Methods

2.1. Microwave Plasma-Enhanced CVD of Nitrogen-Doped Graphene

The direct transfer-less synthesis of nitrogen-doped graphene was performed by employing the microwave plasma-enhanced CVD system Cyrannus (Innovative Plasma Systems (Iplas) GmbH, Troisdorf, Germany). Monocrystalline Si(1 0 0) (UniversityWafer Inc., South Boston, MA, USA) was used as a substrate. Plasma cleaning (power 1.7 kW, operating pressure 22 mbar, temperature 700 °C, hydrogen flow rate 200 sccm, and process duration 10 min) of the substrate was performed until the heater reached the target temperature. A methane, hydrogen, and nitrogen gas mixture was used for the direct synthesis of nitrogen-doped graphene. Afterwards, methane and nitrogen gas were introduced into the chamber. The growth process was performed in a single step. A steel enclosure on the substrate was used for the elimination of the unwanted direct plasma effects. The technological parameters of the plasma (i.e., power 0.7 kW, operating pressure 20 mbar, temperature 700 °C, and process duration 60 min) and the flow rate of hydrogen (75 sccm) and methane (35 sccm) gases were kept constant, while the flow rate of nitrogen gas was varied in the range of 0–110 sccm. Samples were denoted depending on the flow rate of nitrogen used in the process: N0 (N_2, 0 sccm), N35 (N_2, 35 sccm), N75 (N_2, 75 sccm), or N110 (N_2, 110 sccm).

2.2. Characterization

A NanoWizardIII atomic force microscope (JPK Instruments, Bruker Nano GmbH, Berlin, Germany) was used to conduct experiments at room temperature, with data analyzed using a SurfaceXplorer and JPKSPM Data Processing software (Version spm-4.3.13, JPK Instruments, Bruker Nano GmbH). An ACTA probe (Applied NanoStructures, Inc., Mountain View, CA, USA, specification: cantilever shape—pyramidal; radius of curvature < 10.0 nm and cone angle—20°; calibrated spring constant—54.2 N/m; reflex side coating—Al with thickness of 50 nm $\pm$ 5 nm) was used to acquire AFM topographical images in contact mode.

Raman spectroscopy was performed using an inVia Raman spectrometer (Renishaw, Wotton-under-Edge, UK) equipped with a semiconductor green laser (wavelength of 532 nm), 2400 lines/mm grating, confocal microscope (50× objective), and CCD camera. The 5% laser output power was used for spectra recording (10 × 10 s accumulation time) in order to avoid sample damage. Band deconvolution into separate components was performed with OriginPro 8.0 software (OriginLab, Northampton, MA, USA).

Chemical state changes were investigated by employing a Thermo Scientific ESCALAB 250Xi spectrometer with monochromatized Al Kα radiation (hν = 1486.6 eV): X-ray spot of 0.9 mm in diameter; base pressure better than 3×10^{-9} Torr; 40 eV pass energy in transmission mode; energy scale calibration according to Au $4f_{7/2}$, Ag $3d_{5/2}$ and Cu $2p_{3/2}$. The peak fitting procedure was performed using the original ESCALAB 250Xi Avantage software.

3. Results and Discussion

The AFM analysis of pristine graphene (Figure 1a) was conducted over a 1.1×1.1 μm² area for quantitative morphological evaluation. The pristine graphene surface had a random distribution of surface features (mean height, Z_{mean} of 0.76 ± 0.2 nm) with varying angle orientation to each other, without a preferred direction. The root mean square roughness (R_q) was determined to be 0.16 ± 0.2 nm. The skewness (R_{sk}) was determined to be 0.19 ± 0.2, indicating that the surface peaks dominate over the valleys. The pristine graphene surface exhibited a leptokurtoic distribution of surface features with a kurtosis (R_{ku}) value of 3.32 ± 0.2.

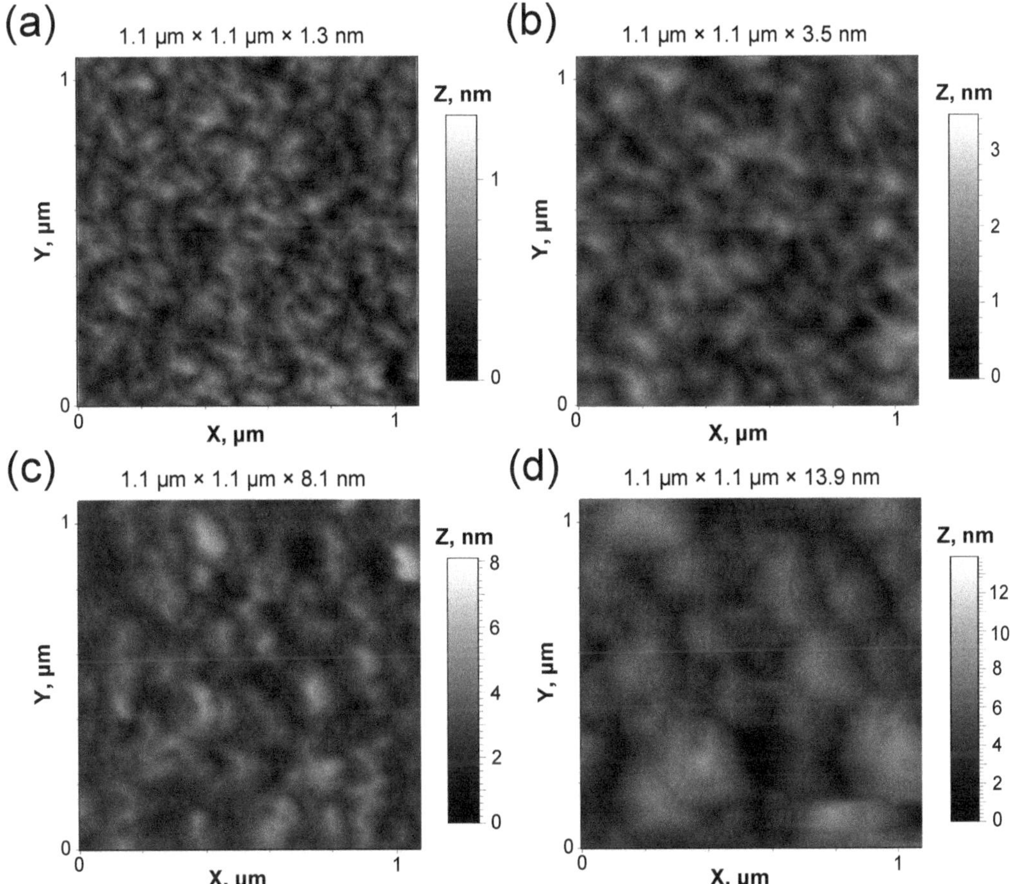

Figure 1. AFM surface topography of pristine graphene (**a**), N35 (**b**), N75 (**c**), and N110 (**d**).

Some significant changes in surface topography (Figure 1b–d) as well as morphology (Table 1) can be observed for nitrogen-doped graphene: with the increase in nitrogen flow rate, the grain-like surface feature size increases in dimensions, with deeper valleys formed in between the boundaries (Figure S1). In general, the height of the surface structures and the roughness increase with the nitrogen flow rate. This change can be explained by the collision between nitrogen and carbon atoms as well as inner atom substitutional placement in the sp^2 hybridized lattice that results in the formation of various nitrogen-induced defect complexes [44], which evidently leads to a more distorted growth process of the graphene structure. Furthermore, graphene synthesis is performed in hydrogen plasma, which is also known to be responsible for defect formation [45,46]. Figure S2 shows the AFM step profiles for the corresponding samples. The thickness of the synthesized graphene films was found to be in the range of $9–13 \pm 1$ nm with no clear dependence on the nitrogen gas flow rate.

Table 1. Surface morphological parameters.

Sample No.	Morphological Parameter			
	Z_{mean}, nm	R_q, nm	R_{sk}	R_{ku}
N0	0.76 ± 0.2	0.16 ± 0.2	0.19 ± 0.2	3.32 ± 0.2
N35	1.65 ± 0.2	0.49 ± 0.2	0.35 ± 0.2	3.09 ± 0.2
N75	3.12 ± 0.2	1.33 ± 0.2	0.93 ± 0.2	4.12 ± 0.2
N110	7.30 ± 0.2	2.32 ± 0.2	0.18 ± 0.2	3.14 ± 0.2

Figure 2 shows Raman spectra of pristine graphene grown directly on monocrystalline Si(1 0 0) substrate, as well as nitrogen-doped graphene samples. Raman spectra are presented in two actual wavenumber ranges of $1100–1800$ cm^{-1} and $2450–3150$ cm^{-1}, respectively. Four deconvoluted bands were determined using Gaussian components consisting of D and G bands at 1347 cm^{-1} and 1599 cm^{-1} and 2D and G + D bands at 2687 cm^{-1} and 2937 cm^{-1} for pristine graphene. The high I_D/I_G ratio of 1.26 indicates highly defective graphene. The low I_{2D}/I_G (0.15) and the broad 2D band in the Raman spectrum indicates the multilayer structure of graphene. The nitrogen doping of graphene is confirmed by the increased I_D/I_G ratio of samples N35 and N75 (Figure 2b,c), as well as the blue shift of the G band position at 1596 and 1595 cm^{-1} for N75 and N110 (Figure 2c,d), respectively. Additionally, a red shift of the 2D position is observed for N110. Previously, Raman studies of nitrogen-doped graphene were conducted in depth to explain the red and blue shift behavior of 2D and G peak positions, corresponding to n-type doping and compressive/tensile strain in graphene [38,44,47]. S. Zheng et al. investigated the metal-catalyst-free growth of graphene on insulating substrates via ammonia-assisted microwave plasma-enhanced CVD [48]. They also observed the blue shift of the G peak position for nitrogen-doped graphene samples.

XPS was employed for the investigation of chemical state changes in the graphene structure resulting from the nitrogen doping. Figure S3 shows high-resolution XPS spectra in C 1s region for pristine graphene, N35 and N110, respectively. The deconvolution of spectra for these samples showed similar results: the strongest component at 284.2 eV, a less intense component at 284.8 eV, and a low-intensity component at 285.7 eV. A highly asymmetric component at 284.2 eV was assigned to C–C (sp^2) bonds [22,49]. The second component at 284.6 eV was assigned to C–C (sp^3) bonds [22,50]. The low-intensity component at 285.7 eV was assigned to C–O–C bonds [51,52]. It was found that the sp^2/sp^3 ratio decreased with nitrogen gas flow rate (Figure S3). Figure 3a does not indicate nitrogen-related peaks in the high-resolution N 1s spectrum, confirming the pristine nature of graphene. Figure 3b,c show the high-resolution XPS N 1s spectra with deconvoluted components of N35 and N110. The deconvoluted components in the high-resolution XPS N 1s spectra were attributed to pyridinic N (398.9 eV and 398.6 eV) [53,54], sp^2 C–N bonds (398.0 eV and 397.8 eV) [55,56], and N–Si bonds (397.2 eV and 397.0 eV) [57,58]. The latter comes from the contribution of the substrate as nitrogen-doped graphene was grown directly on monocrystalline Si(1 0 0). It is also important to note that the sp^2 C–N bonding configuration is similar to pyridine. On the basis of these results, it was determined that the direct microwave plasma-enhanced CVD process produced graphene doped with pure pyridinic N. It was previously reported that the pyridinic N dopant efficiently changes the structure of the graphene valence band, including increasing the density of π states near the Fermi level, as well as reducing work function [59]. Lower work function can dramatically enhance the emitting current in graphene-based electronic devices [60]. It was also demonstrated in [61] that pyridinic nitrogen configuration in graphene contributed to the high catalytic performance. D. Wei et al. investigated the low-temperature critical growth of nitrogen-doped graphene on dielectrics via plasma-enhanced CVD [62]. They also found that nitrogen atoms in graphene are mainly bonded in the pyridinic N form (i.e., a nitrogen atom with two carbon atom neighbors assembling a hexagonal ring). The results of the atomic concentration calculation (Table 2) showed an increase in nitrogen from 1.04 at.% to 2.08 at.% for N35 and N110, respectively, indicating an increased level of pyridinic N doping with an increase in nitrogen gas flow rate. A very similar nitrogen concentration (i.e., 2.0 at.%) was reported for nitrogen-doped graphene synthesized using the N$_2$:CH$_4$ ratio of 2:1 via the direct microwave plasma-enhanced CVD process [44]. A nitrogen concentration of 2.0 at.% was also reported for nitrogen-doped graphene synthesized using a camphor:melamine ratio of 1:3 via the atmospheric pressure CVD process [38].

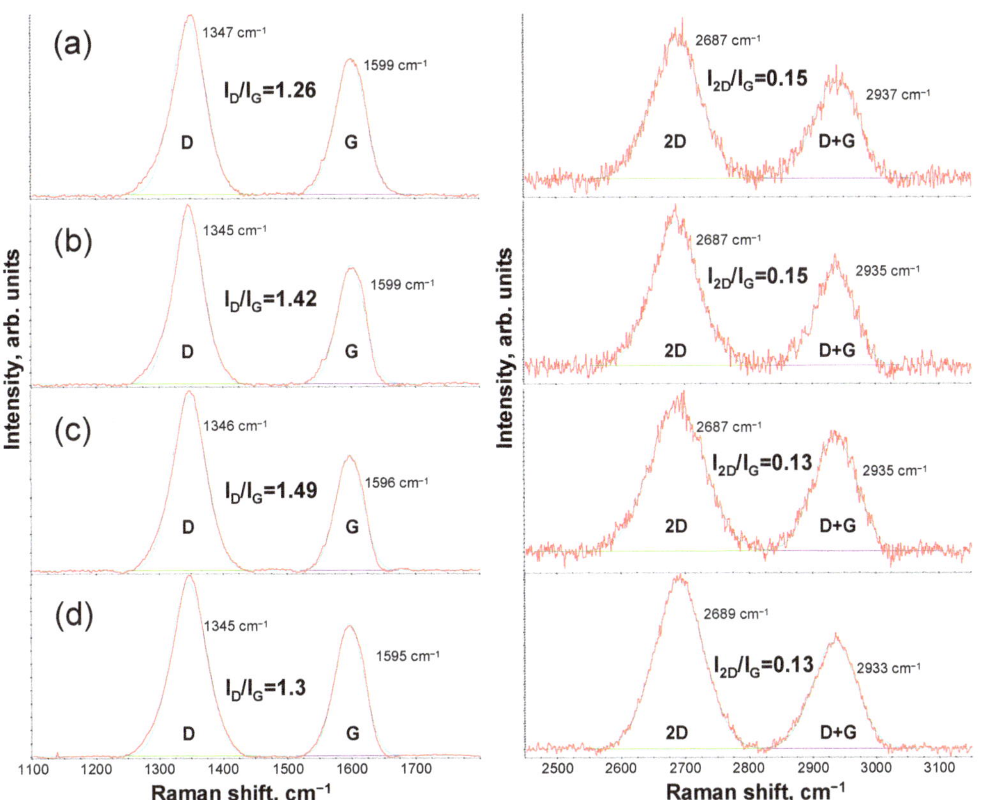

Figure 2. Raman spectra of pristine graphene (**a**), N35 (**b**), N75 (**c**), and N110 (**d**) recorded at 532 nm excitation wavelength and presented in two actual wavenumber ranges.

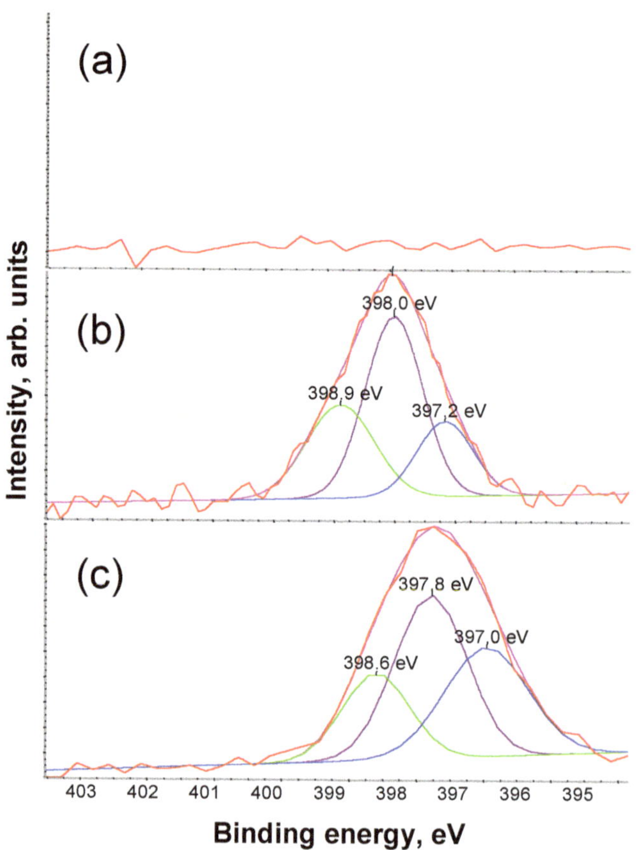

Figure 3. Deconvoluted high-resolution XPS spectra in N 1s region of pristine graphene (**a**), N35 (**b**), and N110 (**c**).

Table 2. Atomic concentration calculation results.

Peak	Atomic Concentration (%)		
	N0	**N35**	**N100**
O 1s	18.48	21.25	21.63
N 1s	0	1.04	2.08
C 1s	47.69	36.37	36.29
Si 2p	33.83	41.35	40

4. Conclusions

We demonstrated the catalyst-less and transfer-less synthesis of nitrogen-doped graphene in a single step at a reduced temperature of 700 °C by using direct microwave plasma-enhanced CVD. The nitrogen flow rate was varied to explore the structural and chemical peculiarities of nitrogen-doped graphene. AFM analysis showed that the height of the graphene surface structures and the roughness increase with the nitrogen flow rate due to nitrogen-induced defect complexes, which evidently lead to a more distorted growth process of the planar graphene. These structural changes were also quantified via Raman spectroscopy. Furthermore, it was determined that nitrogen-doped multilayer graphene structures were produced using this method. XPS analysis showed that pure pyridinic

N was incorporated into the graphene structure during the simultaneous doping process. The level of pyridinic N doping increased with the nitrogen gas flow rate. Nitrogen-doped graphene could have potential applications in optoelectronics.

Supplementary Materials: The following are available online at https://www.mdpi.com/article/10.3390/coatings12050572/s1. Figure S1: Characteristic height profiles for the corresponding lines drawn in AFM images of pristine graphene (a), N35 (b), N75 (c), and N110 (d), Figure S2: Characteristic step profiles for the corresponding lines drawn in AFM images of pristine graphene (a), N35 (b), N75 (c), and N110 (d), Figure S3: Deconvoluted high-resolution XPS spectra in C 1s region of N0, N35, and N110.

Author Contributions: Conceptualization, R.G., Š.M. and A.L.; investigation, R.G., A.L., Š.M. and M.A.; writing—original draft preparation, Š.M. and A.L.; writing—review and editing, A.L. and Š.M.; visualization, A.L.; project administration, Š.M.; funding acquisition, Š.M. All authors have read and agreed to the published version of the manuscript.

Funding: This research was (and is) funded by the European Social Fund under the No. 09.3.3-LMT-K-712-01 "Improvement of researchers' qualification by implementing world-class R&D projects" measure. Grant No. 09.3.3-LMT-K-712-01-0183.

Institutional Review Board Statement: Not applicable.

Informed Consent Statement: Not applicable.

Data Availability Statement: Not applicable.

Acknowledgments: A special thanks go to A. Vasiliauskas, A. Guobienė, K. Šlapikas, V. Stankus, D. Peckus, E. Rajackaitė, T. Tamulevičius, A. Jurkevičiūtė, Š. Jankauskas, and F. Kalyk for technical assistance.

Conflicts of Interest: The authors declare no conflict of interest.

References

1. Mas-Balleste, R.; Gomez-Navarro, C.; Gomez-Herrero, J.; Zamora, F. 2D materials: To graphene and beyond. *Nanoscale* **2011**, *3*, 20–30. [CrossRef] [PubMed]
2. Maniadi, A.; Vamvakaki, M.; Suchea, M.; Tudose, I.V.; Popescu, M.; Romanitan, C.; Pachiu, C.; Ionescu, O.N.; Viskadourakis, Z.; Kenanakis, G. Effect of graphene nanoplatelets on the structure, the morphology, and the dielectric behavior of low-density polyethylene nanocomposites. *Materials* **2020**, *13*, 4776. [CrossRef] [PubMed]
3. Wang, Q.H.; Kalantar-Zadeh, K.; Kis, A.; Coleman, J.N.; Strano, M.S. Electronics and optoelectronics of two-dimensional transition metal dichalcogenides. *Nat. Nanotechnol.* **2012**, *7*, 699–712. [CrossRef] [PubMed]
4. Poot, M.; van der Zant, H.S. Nanomechanical properties of few-layer graphene membranes. *Appl. Phys. Lett.* **2008**, *92*, 063111. [CrossRef]
5. Butler, S.Z.; Hollen, S.M.; Cao, L.; Cui, Y.; Gupta, J.A.; Gutiérrez, H.R.; Heinz, T.F.; Hong, S.S.; Huang, J.; Ismach, A.F. Progress, challenges, and opportunities in two-dimensional materials beyond graphene. *ACS Nano* **2013**, *7*, 2898–2926. [CrossRef]
6. Sun, P.; Wang, K.; Zhu, H. Recent developments in graphene-based membranes: Structure, mass-transport mechanism and potential applications. *Adv. Mater.* **2016**, *28*, 2287–2310. [CrossRef]
7. Akinwande, D.; Brennan, C.J.; Bunch, J.S.; Egberts, P.; Felts, J.R.; Gao, H.; Huang, R.; Kim, J.-S.; Li, T.; Li, Y. A review on mechanics and mechanical properties of 2D materials—Graphene and beyond. *Extrem. Mech. Lett.* **2017**, *13*, 42–77. [CrossRef]
8. Maiti, U.N.; Lee, W.J.; Lee, J.M.; Oh, Y.; Kim, J.Y.; Kim, J.E.; Shim, J.; Han, T.H.; Kim, S.O. 25th anniversary article: Chemically modified/doped carbon nanotubes & graphene for optimized nanostructures & nanodevices. *Adv. Mater.* **2014**, *26*, 40–67.
9. Lv, R.; Terrones, M. Towards new graphene materials: Doped graphene sheets and nanoribbons. *Mater. Lett.* **2012**, *78*, 209–218. [CrossRef]
10. Georgakilas, V.; Otyepka, M.; Bourlinos, A.B.; Chandra, V.; Kim, N.; Kemp, K.C.; Hobza, P.; Zboril, R.; Kim, K.S. Functionalization of graphene: Covalent and non-covalent approaches, derivatives and applications. *Chem. Rev.* **2012**, *112*, 6156–6214. [CrossRef]
11. Dong, X.; Long, Q.; Wei, A.; Zhang, W.; Li, L.-J.; Chen, P.; Huang, W. The electrical properties of graphene modified by bromophenyl groups derived from a diazonium compound. *Carbon* **2012**, *50*, 1517–1522. [CrossRef]
12. Tripathi, N.; Pavelyev, V.; Sharma, P.; Kumar, S.; Rymzhina, A.; Mishra, P. Review of titanium trisulfide (TiS3): A novel material for next generation electronic and optical devices. *Mater. Sci. Semicond. Process.* **2021**, *127*, 105699. [CrossRef]
13. Wei, D.; Liu, Y.; Wang, Y.; Zhang, H.; Huang, L.; Yu, G. Synthesis of N-doped graphene by chemical vapor deposition and its electrical properties. *Nano Lett.* **2009**, *9*, 1752–1758. [CrossRef]
14. Peres, N.; Guinea, F.; Neto, A.C. Coulomb interactions and ferromagnetism in pure and doped graphene. *Phys. Rev. B* **2005**, *72*, 174406. [CrossRef]

15. Uchoa, B.; Neto, A.C. Superconducting states of pure and doped graphene. *Phys. Rev. Lett.* **2007**, *98*, 146801. [CrossRef] [PubMed]
16. Wang, X.; Sun, G.; Routh, P.; Kim, D.-H.; Huang, W.; Chen, P. Heteroatom-doped graphene materials: Syntheses, properties and applications. *Chem. Soc. Rev.* **2014**, *43*, 7067–7098. [CrossRef] [PubMed]
17. Wu, P.; Cai, Z.; Gao, Y.; Zhang, H.; Cai, C. Enhancing the electrochemical reduction of hydrogen peroxide based on nitrogen-doped graphene for measurement of its releasing process from living cells. *Chem. Commun.* **2011**, *47*, 11327–11329. [CrossRef]
18. Li, N.; Wang, Z.; Zhao, K.; Shi, Z.; Gu, Z.; Xu, S. Large scale synthesis of N-doped multi-layered graphene sheets by simple arc-discharge method. *Carbon* **2010**, *48*, 255–259. [CrossRef]
19. Jeong, H.M.; Lee, J.W.; Shin, W.H.; Choi, Y.J.; Shin, H.J.; Kang, J.K.; Choi, J.W. Nitrogen-doped graphene for high-performance ultracapacitors and the importance of nitrogen-doped sites at basal planes. *Nano Lett.* **2011**, *11*, 2472–2477. [CrossRef]
20. Ion-Ebrașu, D.; Andrei, R.D.; Enache, S.; Căprărescu, S.; Negrilă, C.C.; Jianu, C.; Enache, A.; Boerașu, I.; Carcadea, E.; Varlam, M. Nitrogen functionalization of cvd grown three-dimensional graphene foam for hydrogen evolution reactions in alkaline media. *Materials* **2021**, *14*, 4952. [CrossRef]
21. Kamedulski, P.; Truszkowski, S.; Lukaszewicz, J.P. Highly effective methods of obtaining N-doped graphene by gamma irradiation. *Materials* **2020**, *13*, 4975. [CrossRef] [PubMed]
22. Komissarov, I.V.; Kovalchuk, N.G.; Labunov, V.A.; Girel, K.V.; Korolik, O.V.; Tivanov, M.S.; Lazauskas, A.; Andrulevičius, M.; Tamulevičius, T.; Grigaliūnas, V. Nitrogen-doped twisted graphene grown on copper by atmospheric pressure CVD from a decane precursor. *Beilstein J. Nanotechnol.* **2017**, *8*, 145–158. [CrossRef] [PubMed]
23. Lü, X.; Wu, J.; Lin, T.; Wan, D.; Huang, F.; Xie, X.; Jiang, M. Low-temperature rapid synthesis of high-quality pristine or boron-doped graphene via Wurtz-type reductive coupling reaction. *J. Mater. Chem.* **2011**, *21*, 10685–10689. [CrossRef]
24. Jeon, I.-Y.; Choi, H.-J.; Ju, M.J.; Choi, I.T.; Lim, K.; Ko, J.; Kim, H.K.; Kim, J.C.; Lee, J.-J.; Shin, D. Direct nitrogen fixation at the edges of graphene nanoplatelets as efficient electrocatalysts for energy conversion. *Sci. Rep.* **2013**, *3*, 2260. [CrossRef]
25. Rani, P.; Jindal, V. Designing band gap of graphene by B and N dopant atoms. *RSC Adv.* **2013**, *3*, 802–812. [CrossRef]
26. Lherbier, A.; Botello-Méndez, A.R.; Charlier, J.-C. Electronic and transport properties of unbalanced sublattice N-doping in graphene. *Nano Lett.* **2013**, *13*, 1446–1450. [CrossRef]
27. Liu, Z.W.; Peng, F.; Wang, H.J.; Yu, H.; Zheng, W.X.; Yang, J. Phosphorus-doped graphite layers with high electrocatalytic activity for the O_2 reduction in an alkaline medium. *Angew. Chem. Int. Ed.* **2011**, *50*, 3257–3261. [CrossRef]
28. Denis, P.A. Band gap opening of monolayer and bilayer graphene doped with aluminium, silicon, phosphorus, and sulfur. *Chem. Phys. Lett.* **2010**, *492*, 251–257. [CrossRef]
29. Ribas, M.A.; Singh, A.K.; Sorokin, P.B.; Yakobson, B.I. Patterning nanoroads and quantum dots on fluorinated graphene. *Nano Res.* **2011**, *4*, 143–152. [CrossRef]
30. Yang, M.; Zhou, L.; Wang, J.; Liu, Z.; Liu, Z. Evolutionary chlorination of graphene: From charge-transfer complex to covalent bonding and nonbonding. *J. Phys. Chem. C* **2012**, *116*, 844–850. [CrossRef]
31. Venkatesan, N.; Archana, K.S.; Suresh, S.; Aswathy, R.; Ulaganthan, M.; Periasamy, P.; Ragupathy, P. Boron-Doped Graphene as Efficient Electrocatalyst for Zinc-Bromine Redox Flow Batteries. *ChemElectroChem* **2019**, *6*, 1107–1114. [CrossRef]
32. Kalita, G.; Wakita, K.; Takahashi, M.; Umeno, M. Iodine doping in solid precursor-based CVD growth graphene film. *J. Mater. Chem.* **2011**, *21*, 15209–15213. [CrossRef]
33. Yu, X.; Li, N.; Zhang, S.; Liu, C.; Chen, L.; Xi, M.; Song, Y.; Ali, S.; Iqbal, O.; Han, M. Enhancing the energy storage capacity of graphene supercapacitors via solar heating. *J. Mater. Chem. A* **2022**, *10*, 3382–3392. [CrossRef]
34. Chen, J.; Bailey, J.J.; Britnell, L.; Perez-Page, M.; Sahoo, M.; Zhang, Z.; Strudwick, A.; Hack, J.; Guo, Z.; Ji, Z. The performance and durability of high-temperature proton exchange membrane fuel cells enhanced by single-layer graphene. *Nano Energy* **2022**, *93*, 106829. [CrossRef]
35. Mu, Y.; Han, M.; Li, J.; Liang, J.; Yu, J. Growing vertical graphene sheets on natural graphite for fast charging lithium-ion batteries. *Carbon* **2021**, *173*, 477–484. [CrossRef]
36. Safie, N.E.; Azam, M.A.; Aziz, M.F.; Ismail, M. Recent progress of graphene-based materials for efficient charge transfer and device performance stability in perovskite solar cells. *Int. J. Energy Res.* **2021**, *45*, 1347–1374. [CrossRef]
37. Xie, T.; Wang, Q.; Wallace, R.M.; Gong, C. Understanding and optimization of graphene gas sensors. *Appl. Phys. Lett.* **2021**, *119*, 013104. [CrossRef]
38. Shinde, S.M.; Kano, E.; Kalita, G.; Takeguchi, M.; Hashimoto, A.; Tanemura, M. Grain structures of nitrogen-doped graphene synthesized by solid source-based chemical vapor deposition. *Carbon* **2016**, *96*, 448–453. [CrossRef]
39. Mattevi, C.; Kim, H.; Chhowalla, M. A review of chemical vapour deposition of graphene on copper. *J. Mater. Chem.* **2011**, *21*, 3324–3334. [CrossRef]
40. Huang, L.; Chang, Q.; Guo, G.; Liu, Y.; Xie, Y.; Wang, T.; Ling, B.; Yang, H. Synthesis of high-quality graphene films on nickel foils by rapid thermal chemical vapor deposition. *Carbon* **2012**, *50*, 551–556. [CrossRef]
41. Kang, J.; Shin, D.; Bae, S.; Hong, B.H. Graphene transfer: Key for applications. *Nanoscale* **2012**, *4*, 5527–5537. [CrossRef] [PubMed]
42. Hong, S.K.; Song, S.M.; Sul, O.; Cho, B.J. Carboxylic group as the origin of electrical performance degradation during the transfer process of CVD growth graphene. *J. Electrochem. Soc.* **2012**, *159*, K107. [CrossRef]
43. Ambrosi, A.; Pumera, M. The CVD graphene transfer procedure introduces metallic impurities which alter the graphene electrochemical properties. *Nanoscale* **2014**, *6*, 472–476. [CrossRef] [PubMed]

44. Boas, C.; Focassio, B.; Marinho, E.; Larrude, D.; Salvadori, M.; Leão, C.R.; Dos Santos, D.J. Characterization of nitrogen doped graphene bilayers synthesized by fast, low temperature microwave plasma-enhanced chemical vapour deposition. *Sci. Rep.* **2019**, *9*, 13715. [CrossRef]

45. Kim, J.H.; Castro, E.J.D.; Hwang, Y.G.; Lee, C.H. Synthesis of Few-Layer Graphene Using DC PE-CVD. *AIP Conf. Proc.* **2011**, *1399*, 801–802.

46. Childres, I.; Jauregui, L.A.; Park, W.; Cao, H.; Chen, Y.P. Raman spectroscopy of graphene and related materials. *New Dev. Photon Mater. Res.* **2013**, *1*, 1–20.

47. Zafar, Z.; Ni, Z.H.; Wu, X.; Shi, Z.X.; Nan, H.Y.; Bai, J.; Sun, L.T. Evolution of Raman spectra in nitrogen doped graphene. *Carbon* **2013**, *61*, 57–62. [CrossRef]

48. Zheng, S.; Zhong, G.; Wu, X.; D'Arsiè, L.; Robertson, J. Metal-catalyst-free growth of graphene on insulating substrates by ammonia-assisted microwave plasma-enhanced chemical vapor deposition. *RSC Adv.* **2017**, *7*, 33185–33193. [CrossRef]

49. Wang, C.; Yuen, M.F.; Ng, T.W.; Jha, S.K.; Lu, Z.; Kwok, S.Y.; Wong, T.L.; Yang, X.; Lee, C.S.; Lee, S.T. Plasma-assisted growth and nitrogen doping of graphene films. *Appl. Phys. Lett.* **2012**, *100*, 253107. [CrossRef]

50. Merel, P.; Tabbal, M.; Chaker, M.; Moisa, S.; Margot, J. Direct evaluation of the sp3 content in diamond-like-carbon films by XPS. *Appl. Surf. Sci.* **1998**, *136*, 105–110. [CrossRef]

51. Marinoiu, A.; Raceanu, M.; Andrulevicius, M.; Tamuleviciene, A.; Tamulevicius, T.; Nica, S.; Bala, D.; Varlam, M. Low-cost preparation method of well dispersed gold nanoparticles on reduced graphene oxide and electrocatalytic stability in PEM fuel cell. *Arab. J. Chem.* **2020**, *13*, 3585–3600. [CrossRef]

52. Jankauskaitȿ, V.; Vitkauskienȿ, A.; Lazauskas, A.; Baltrusaitis, J.; Prosyȿevas, I.; Andruleviȿius, M. Bactericidal effect of graphene oxide/Cu/Ag nanoderivatives against Escherichia coli, Pseudomonas aeruginosa, Klebsiella pneumoniae, Staphylococcus aureus and Methicillin-resistant Staphylococcus aureus. *Int. J. Pharm.* **2016**, *511*, 90–97. [CrossRef] [PubMed]

53. He, H.; Yang, Q.; Xiao, S.; Han, X.; Li, Q.; Lv, K.; Hong, J.; Tang, D.; Kejian, D. Investigation of Edge-selectively Nitrogen-doped Metal Free Graphene For Oxygen Reduction Reaction. *J. Adv. Nanotechnol.* **2020**, *1*, 5–13.

54. Tian, K.; Wang, J.; Cao, L.; Yang, W.; Guo, W.; Liu, S.; Li, W.; Wang, F.; Li, X.; Xu, Z. Single-site pyrrolic-nitrogen-doped sp2-hybridized carbon materials and their pseudocapacitance. *Nat. Commun.* **2020**, *11*, 3884. [CrossRef]

55. Liu, F.; Zhang, Z.; Rong, X.; Yu, Y.; Wang, T.; Sheng, B.; Wei, J.; Zhou, S.; Yang, X.; Xu, F. Graphene-Assisted Epitaxy of Nitrogen Lattice Polarity GaN Films on Non-Polar Sapphire Substrates for Green Light Emitting Diodes. *Adv. Funct. Mater.* **2020**, *30*, 2001283. [CrossRef]

56. Wu, Z.; Wang, Y.; Li, S.; Wang, X.; Xu, Z.; Zhou, F. Mechanical and tribological properties of BCN coatings sliding against different wood balls. *Sci. Eng. Compos. Mater.* **2019**, *26*, 402–411. [CrossRef]

57. Zhou, J.; Ye, F.; Cui, X.; Cheng, L.; Li, J.; Liu, Y.; Zhang, L. Mechanical and dielectric properties of two types of Si3N4 fibers annealed at elevated temperatures. *Materials* **2018**, *11*, 1498. [CrossRef]

58. Tao, F.; Wang, Z.H.; Qiao, M.H.; Liu, Q.; Sim, W.S.; Xu, G.Q. Covalent attachment of acetonitrile on Si (100) through Si–C and Si–N linkages. *J. Chem. Phys.* **2001**, *115*, 8563–8569. [CrossRef]

59. Luo, Z.; Lim, S.; Tian, Z.; Shang, J.; Lai, L.; MacDonald, B.; Fu, C.; Shen, Z.; Yu, T.; Lin, J. Pyridinic N doped graphene: Synthesis, electronic structure, and electrocatalytic property. *J. Mater. Chem.* **2011**, *21*, 8038–8044. [CrossRef]

60. Legesse, M.; El Mellouhi, F.; Bentria, E.T.; Madjet, M.E.; Fisher, T.S.; Kais, S.; Alharbi, F.H. Reduced work function of graphene by metal adatoms. *Appl. Surf. Sci.* **2017**, *394*, 98–107. [CrossRef]

61. Sheng, Z.-H.; Shao, L.; Chen, J.-J.; Bao, W.-J.; Wang, F.-B.; Xia, X.-H. Catalyst-free synthesis of nitrogen-doped graphene via thermal annealing graphite oxide with melamine and its excellent electrocatalysis. *ACS Nano* **2011**, *5*, 4350–4358. [CrossRef] [PubMed]

62. Wei, D.; Peng, L.; Li, M.; Mao, H.; Niu, T.; Han, C.; Chen, W.; Wee, A.T.S. Low temperature critical growth of high quality nitrogen doped graphene on dielectrics by plasma-enhanced chemical vapor deposition. *ACS Nano* **2015**, *9*, 164–171. [CrossRef] [PubMed]

Article

Electrical Method for In Vivo Testing of Exhalation Sensors Based on Natural Clinoptilolite

Gianfranco Carotenuto [1,*] and Luigi Nicolais [2]

[1] Institute for Polymers, Composites and Biomaterials (IPCB-CNR), National Research Council, Piazzale E. Fermi, 80055 Naples, Italy

[2] Engineering Faculty, University of Naples "Federico II", Corso Nicolangelo Protopisani, 80146 Naples, Italy; nicolais@unina.it

* Correspondence: giancaro@unina.it; Tel.: +39-3384045292

Citation: Carotenuto, G.; Nicolais, L. Electrical Method for In Vivo Testing of Exhalation Sensors Based on Natural Clinoptilolite. *Coatings* **2022**, *12*, 377. https://doi.org/10.3390/coatings12030377

Academic Editor: María Dolores Fernández Ramos

Received: 21 February 2022
Accepted: 10 March 2022
Published: 13 March 2022

Publisher's Note: MDPI stays neutral with regard to jurisdictional claims in published maps and institutional affiliations.

Abstract: Natural substances with a complex chemical structure can be advantageously used for functional applications. Such functional materials can be found both in the mineral and biological worlds. Owing to the presence of ionic charge carriers (i.e., extra-framework cations) in their crystal lattice, whose mobility is strictly depending on parameters of the external environment (e.g., temperature, humidity, presence of small gaseous polar molecules, etc.), zeolites can be industrially exploited as a novel functional material class with great potentialities in sensors and electric/electronic field. For fast-responding chemical-sensing applications, ionic transport at the zeolite surface is much more useful than bulk-transport, since molecular transport in the channel network takes place by a very slow diffusion mechanism. The environmental dependence of electrical conductivity of common natural zeolites characterized by an aluminous nature (e.g., chabasite, clinoptilolite, etc.) can be conveniently exploited to fabricate impedimetric water-vapor sensors for apnea syndrome monitoring. The high mechanical, thermal, and chemical stability of geomorphic clinoptilolite (the most widely spread natural zeolite type) makes this type of zeolite the most adequate mineral substance to fabricate self-supporting impedimetric water-vapor sensors. In the development of devices for medical monitoring (e.g., apnea-syndrome monitors), it is very important to combine these inexpensive nature-made sensors with a low-weight simplified electronic circuitry that can be easily integrated in wearable items (e.g., garments, wristwatch, etc.). Very low power square-wave voltage sources (micro-Watt voltage sources) show significant voltage drops under only a minimal electric load, and this property of the ac generator can be advantageously exploited for detecting the small impedimetric change observed in clinoptilolite sensors during their exposition to water vapor coming from the human respiratory exhalation. Owing to the ionic conduction mechanism (single-charge carrier) characterizing the zeolite slab surface, the sensor biasing by an ac signal is strictly required. Cheap handheld multimeters frequently include a very low power square-wave (or sinusoidal) voltage source of different frequency (typically 50 Hz or 1 kHz) that is used as a signal injector (signal tracer) to test audio amplifiers (low-frequency amplifies), tone control (equalizer), radios, etc. Such multimeter outputs can be connected in parallel with a true-RMS (Root-Mean-Square) ac voltmeter to detect the response of the clinoptilolite-based impedimetric sensors as voltage drop. The frequency of exhalation during breathing can be measured, and the exhalation behavior can be visualized, too, by using the voltmeter readings. Many handheld multimeters also include a data-logging possibility, which is extremely useful to record the voltage reading over time, thus giving a time-resolved voltage measurement that contains all information concerning the breathing test. Based on the same principle (i.e., voltage drop under minimal resistive load) a devoted electronic circuitry can also be made.

Keywords: clinoptilolite; impedimetric sensor; surface conductivity; apnea syndrome monitoring; voltage drop

1. Introduction

Materials with a unique combination of properties (multifunctional materials) can be found among natural substances characterized by significant structural complexity [1]. When a possible application for such nature-made materials has been identified, the development of a device with performance comparable to that of systems based on artificial substances can follow [2]. There are also examples in the literature in which similar technological results cannot be reached with comparable performance by using manmade materials [3,4]. As a consequence, the capability to identify the applicative potentiality hidden in the most common natural substances represents the key factor to come in the technological exploitation of nature-made materials, thus promoting sustainability. Among the different available natural substances, the most powerful potentialities have been found in complex types of solids, since their artificial replication is quite difficult [5,6]. Numerous natural systems have extreme complexity, because complexity is the intrinsic characteristic of the whole natural world [7]. For example, there are a number of biological tissues, mineral substances, and biominerals with specific physical properties, such as piezoelectricity, electrical conductivity, magnetism, etc. (e.g., onion skin, diatomite, eggshell, etc.) [1,2], that are advantageously exploitable for different technological applications.

Natural zeolites represent an interesting class of multifunctional materials, and a number of potentially useful physical and chemical characteristics of these mineral substances have been industrially exploited [8–14]. Usually, natural zeolites have a quite low Si/Al atomic ratio, and for such a reason they are named "aluminous zeolites", in contrast with the "high-siliceous zeolites" that are mainly originated by synthesis. Owing to the aluminum atoms contained in these chemical compounds, the presence of as many extra-lattice cations required to balance the negative charge localized on the aluminum atom follows. Natural zeolites contain different types of cations (typically Na^+, K^+, Ca^{2+}, Mg^{2+}, and Fe^{3+}), and since these cations may have single or multiple charge, there is a significant fraction of the aluminum atoms in the lattice with unbalanced negative charge. Consequently, electrical transport in zeolite lattice becomes possible by cation hopping among the neighbor negatively charged sites [15,16]. Natural zeolites are therefore solid-state ionic conductors with single-charge carrier (usually corresponding to the largest monovalent cations present in crystal lattice, since they have the lowest charge density [17]). However, these crystalline solids may conduct electricity only at high temperature (since in this case a larger fraction of charge carriers have enough energy exceeding the activation barrier required for jumping between the nearest negative sites) or by hydration/solvation (because in this case the Coulomb interaction between cations and the negative sites decreases) [18–21]. Owing to the presence of a regular array of few angstrom-sized channels (1–13 Å) crossing the zeolite crystal lattice, all cations in the lattice can be solvated only by small polar molecules, that is, molecules with an average size lower than the channel diameter; these include, for example, molecules of water, methanol, formaldehyde, etc. Consequently, natural zeolite crystals can be technologically exploited as both thermal and chemical sensors [22]. In their use as chemical sensor, these materials are able to promptly modify their external surface electrical conductivity by exposition to water vapor. The surface electrical conductivity of natural zeolites with Si/Al close to 5 (e.g., clinoptilolite) may change significantly by exposition to water vapor, and in addition the water molecules spontaneously desorb from these slightly hydrophobic materials when these last are removed from the humid environment. Therefore, zeolite chemical sensors show completely reversible electrical behavior with fast return to the original electrical conductivity value. Owing to the involved ionic conduction mechanism, impedance measurements are required for zeolitic sensors. Sinusoidal or square-wave signals with frequency lower than 1 kHz can be conveniently used. A slab of natural zeolite with a convenient value of the Si/Al ratio can be used as a gas-phase water sensor, and, according to the literature [23], geomorphic clinoptilolite has shown to be very adequate for such an application.

Natural clinoptilolite is the most widely spread zeolite type of natural origin [24]. This crystalline substance is characterized by low cost and excellent mechanical, thermal, and chemical resistance. A slab of natural clinoptilolite has a polycrystalline nature made of highly compacted few micron-sized single lamellar crystals [25,26]. Geomorphic clinoptilolite can be readily used to fabricate water-vapor sensors by applying two parallel electrodes to the surface of a small and perfectly flat slab. The electrodes can be simply obtained by painting two very close rectangular areas on the slab surface by using silver paint. Electrodes must be located on the slab surface in order to exclude from the electrical transport mechanism those cations belonging to internal channel surface. Thus, preventing the slow diffusive mass transport mechanism involved in the solvation of inner cations, the maximum response-fastness possible for this ceramic sensor can be achieved. Owing to the excellent mechanical resistance of clinoptilolite, very robust self-supporting sensors can be fabricated. For the high thermal stability of clinoptilolite, these water-vapor sensors can be exploited for technological applications also at temperatures higher than room temperature.

Recently, the use of naturally available nanostructured mineral substances (e.g., halloysite nanotubes, sepiolite nanofibers, attapulgite, etc.) in the fabrication of resistive humidity sensors to be used for different body-related-humidity-detection applications (e.g., respiratory behavior, speech recognition, skin moisture, non-contact switch, diaper monitoring, etc. [27]) has been proposed in the literature [28–30]. These nature-made sensitive materials have shown good humidity-sensing performances at different relative humidity (RH) conditions, and, in addition, their use represents a facile, low-cost, and environmentally friendly strategy to achieve high-performance sensing devices, without requiring complex manufacturing approaches frequently adopted in this field [31–33].

Inexpensive fast-response ceramic moisture sensors can be conveniently used as devices for apnea syndrome control (i.e., human exhalation sensors) [34–36] by using a miniaturized wearable electric revelation technique. Here, the capability of geomorphic clinoptilolite sensor to promptly and reversibly detect the humidity exhaled during breathing was evaluated by using the "voltage-drop" technique, an original instrumental approach that requires very simple electronic circuitry. Device miniaturizing needs extremely reduced electronic circuitry, which is essential for the wearability of these health monitors (e.g., apnea-syndrome monitor).

2. Materials and Methods

A commercial sample of natural clinoptilolite (TIP—Technische Industrie Produkte, GmbH, Waibstadt, Germany) was used in the as-received form for fabricating prototypes of the impedimetric sensor. In particular, the sensors were obtained by cutting the raw zeolite piece in the form of rectangular slabs of small thickness (5.0 mm $\times$ 10.0 mm $\times$ 3.0 mm) by using a mini electric cutting machine (electric mini-grinder, VUM-40, Vigor, Fossano (CN), Italy, equipped with a diamond cutting wheel). Two electrical contacts (1 mm spaced) were painted on the slab surface by using a curable silver paste (EN-06B8, ENSON, Shenzhen, China), and wires were cold-welded to the electrodes by the same paint. According to silver-paste specifications, the obtained device was left to dry in air for 2 days and then baked in an oven at 140 °C for 30 min. Two-lead measurements were used to test the moisture sensor; this type of measurement involves a small contact resistance due to the silver–zeolite interface that slightly reduces the device sensitivity. Since the silver electrodes are located at the zeolite surface, only hydrated monovalent cations (i.e., KOH^{2+}) belonging to external surface can participate in the electrical transport.

The clinoptilolite sample morphology was investigated by scanning electron microscopy (SEM, Quanta 200 FEG microscope, FEI, Hillsboro, OR, USA) after its powdering by a hammer. The characteristic Si/Al molar ratio and the type of extra-framework cations present in the mineral were determined by using energy-dispersive X-ray spectroscopy (EDS, Inca 250, Oxford Instruments, Oxford, UK). The type of crystalline solid phases present inside the mineral were identified by using large-angle powder diffraction (XRD, X'Pert PRO, PANalytical, Oxford, UK).

A handheld digital multimeter (DMM, UT-71D, Uni-Trend, Dongguan, China) was used as a true-RMS voltmeter for measuring the voltage across the ac generator output, when it was connected in parallel to the zeolite-based humidity sensor. The DMM included an optically insulated high-speed data-logging system, which allowed the multimeter to record voltage measurements at a speed of 8–9 Sa/s (sampling per second). Such high-speed voltage recording required setting the multimeter in low-resolution mode (4000 counts). In particular, DMM was connected to a PC by a cable, and measurements were recorded by using a dedicated software (UTC/D/E Interface Program, version 3.00, 2017, Uni-Trend Technology, Dongguan, China). Square-wave voltage sources embedded into different handheld DMMs (i.e., DT830D, DT-830B, DT832, ANENG AN8206, ANENG AN8008, KONIG KDM-100, and UT20B) were tested. The best results, in terms of lowest power source, were found with the square-wave source embedded in the DT830D digital multimeter (cheap entry-level DMM, available under different trademarks). The square-wave trace was obtained by an analog oscilloscope (AO-610-2 10 MHz, Voltcraft, distributed by Conrad Electronic SE, Hirschau, Germany), and the harmonic composition of the square-wave signal was obtained by a Fast Fourier Transform (FFT) analysis based on a 2-channel 10 MHz USB oscilloscope (PicoScope 2204A-D2, Pico Technology, Cambridgeshire, UK). A resistance decade box (3280, PeakTech GmbH, Ahrensburg, Germany) was also used to determine the I–V characteristics of the square-wave sources.

3. Results and Discussion

Natural clinoptilolite usually consists of a combination of different crystalline solid phases (zeolites, quartz, etc.), and the clinoptilolite phase is only the principal component of such a natural composite. As a consequence, the exact composition of the mineral needs to be experimentally determined by using the X-ray diffraction technique (XRD). According to the diffractogram shown in Figure 1, the clinoptilolite sample contained the following main crystalline solid phases: clinoptilolite (48.4 wt.%), anorthite (42.0 wt.%), quartz (8.9 wt.%), and stilbite (0.7 wt.%); indeed, the most intensive peaks belonging to the diffraction patterns of these minerals can be clearly detected in the XRD of Figure 1. According to the literature, the found crystalline phase composition is quite typical for clinoptilolite of natural origin [24].

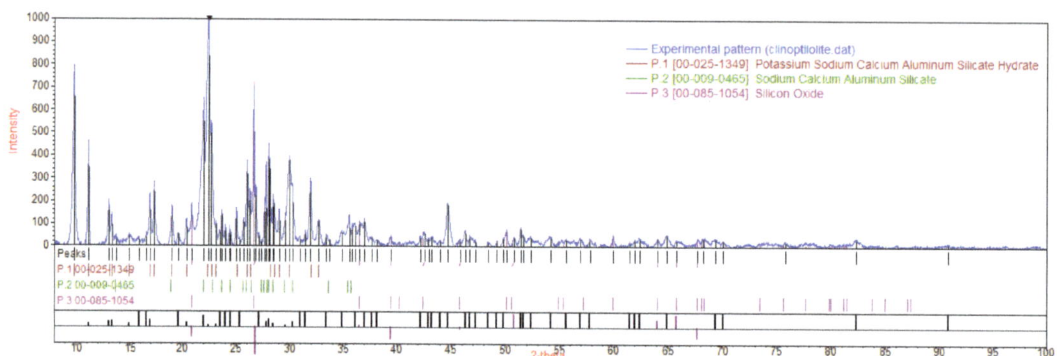

Figure 1. XRD of the natural zeolite used for fabricating the impedimetric water sensor with indication of peaks of the clinoptilolite crystalline phase.

Geomorphic clinoptilolite has a polycrystalline structure. Owing to the extremely high pressure applied to the stone for millions of years, such polycrystalline structure results highly consolidated and characterized by a density (2.15 g/cm^3) close to that of a zeolite single-crystal (absence of macro-porosity) [26]. The clinoptilolite phase is made of perfectly staked single-lamellar crystals whose morphology can be easily visualized by Scanning Electron Microscopy (SEM) after having delaminated the mineral by applying a strong compressive stress (e.g., hammer blow). The lamellar morphology of the clinoptilolite single crystals is shown in Figure 2a. As visible, all single-lamellar crystals have exactly the same thickness, corresponding to 40 nm; while the other two sizes range between 300 nm and 1 μm.

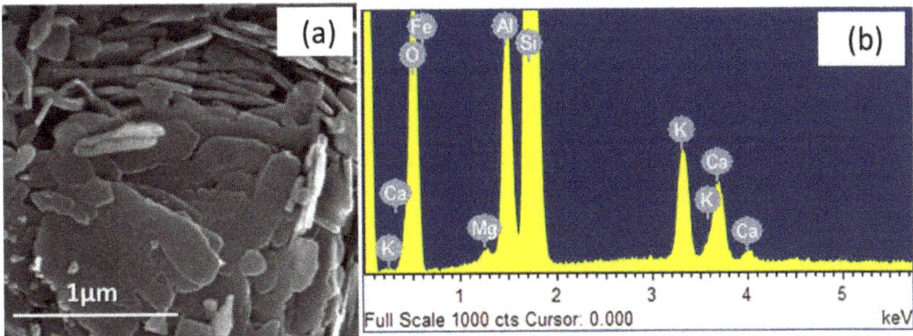

Figure 2. SEM-micrograph of the clinoptilolite single lamellar crystals generated by sample delamination (**a**) and EDS spectrum of the geomorphic clinoptilolite surface (**b**).

The characteristic Si/Al molar ratio and the type of extra-framework cations contained in the sample have been established by Energy-Dispersive Spectroscopy (EDS). The EDS-spectrum of the sample is shown in Figure 2b; the specimen contained the following types of cations: K^+, Ca^{2+}, Mg^{2+}, and Fe^{3+} (in a very little amount). Owing to the low charge density, potassium cations (K^+) were the only charge-carriers active in the transport for this ionic conductor. The Si/Al atomic ratio of the clinoptilolite sample can be approximately evaluated by using the silicon and aluminum intensities of signals in the EDS spectrum, and it corresponded to ca. 5.4. Such a value is typical for this kind of mineral. According to this moderately high Si/Al value, the clinoptilolite sample can be considered as a quite hydrophobic zeolite type. The scarce hydrophilicity of the mineral is a property of fundamental importance for a moisture-sensing material, since it allows the sensor to behave reversibly in service.

Since voltage is an electrical property that is easy to measure, the possibility to use a root-mean-square (RMS) digital voltmeter to detect the wearable sensor response is very convenient. Usually, the response to stimuli of an impedimetric sensor is detected as impedance (Z) variation that is measured by an LCR-meter, or as variation of the intensity of current flowing in the device. However, if the impedimetric sensor is biased by a very low power generator, the sensor response can be detected also as voltage drop at sensor electrodes. In particular, the lower the generator power is, the higher the sensitivity of this method. In addition, the I–V characteristics of the generator can be determined and the voltage response can be converted to current intensity or impedance variation by using the generator I–V characteristics.

Owing to the prompt response and reversible behavior, the moisture sensor based on clinoptilolite can be used as apnea syndrome monitor. The human exhalation pattern could be obtained by measuring the current intensity flowing at sample surface after biasing this sensor, for example, with a 5 kHz sinusoidal voltage signal (20 V_{pp}). However, a similar pattern with much higher resolution can be generated by using the here proposed "voltage-drop" method. To the best of our knowledge, such an instrumental approach, based on a combination of an extremely low power ac generator and true-RMS digital voltmeter, has never been proposed to detect the response of an impedimetric ceramic sensor, such as, for example, a geomorphic zeolite moisture sensor.

When a zeolitic sensor is biased by a sinusoidal or square-wave voltage signal, produced by a standard ac source (function generator with power >1 W), the very high impedance (few MΩ) of the zeolitic sensor can cause just a negligible voltage drop during the exhalation detection by the sensor. However, many digital multimeters (DMMs) incorporate a square-wave source (signal tracer) of very low power (of the μW order), and if these generators are used to bias the sensor, a significant voltage drop can be experienced even for a slight variation of the high load impedance. Such a type of signal generator is adequate to detect stimuli by an impedimetric zeolitic sensor by measuring the voltage drop at the sensor electrodes. An alternated voltage source is required to avoid ion accumulation at electrodes surface. In particular, the square-wave output incorporated in the DT830D multimeter originates from the liquid crystal display (LCD) driver in the 7106 multimeter chip. The integrated 7106 is an analog-to-digital converter (ADC), which also provides the LCD back panel driving signal with flipping polarity.

The zeolite sensor was connected to the square-wave output, and a true-RMS digital voltmeter (UT71D digital multimeter, UNI-Trend Technology, Dongguan, China) was placed in parallel to this generator, according to the electrical scheme shown in Figure 3. The square-wave signal characteristics are show in Figure 4a,b. As visible in the oscilloscope trace, when a coupling capacitor of 10 nF was used to filter the small offset voltage contained in the signal, the power source gave a perfectly symmetrical square-wave with an amplitude of 5.0 V_{pp} (for a square-wave signal, the effective voltage, V_{eff}, as measured by a true-RMS digital voltmeter, exactly corresponds to the peak voltage value, $V_p = 2.5$). As is visible in Figure 4b, the square-wave signal analysis in the frequency domain showed that the signal was composed by the fundamental at the frequency of 50 Hz and three main odd harmonics.

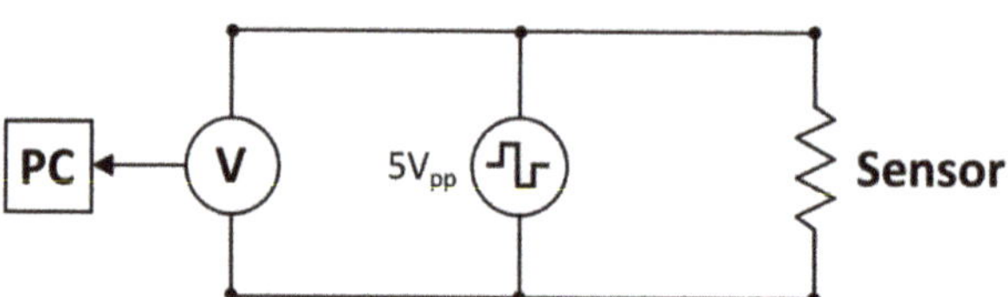

Figure 3. Electrical circuit used to record the voltage-drop measurements.

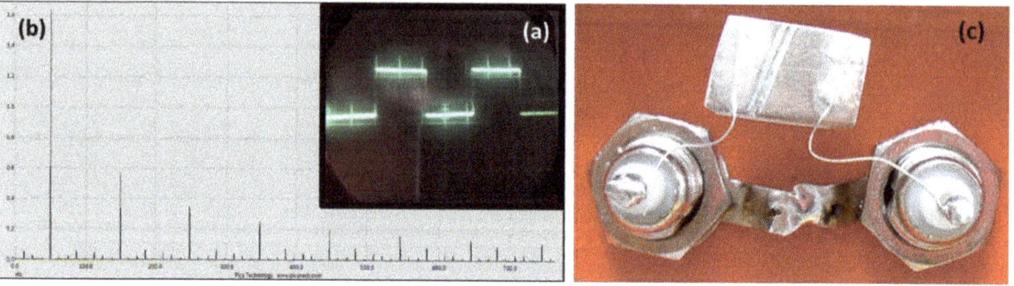

Figure 4. Square-wave signal (**a**) with its harmonic analysis by FFT oscilloscope (**b**) and sensor prototype (**c**).

Time-resolved true-RMS voltage measurements were recorder during exposition to human breathing by using the digital voltmeter UT71D data-logging system. An example of a breathing pattern, including three exhalation steps obtained by the voltage drop technique, is shown in Figure 5. According to this breathing pattern, the surface resistivity was quickly modified in the presence of the water vapor, and the voltage decreased by ca. 100 mV in average. As is visible, water adsorption was a very fast process, and it was described by a linear temporal behavior, while water desorption was slightly slower and followed a parabolic law.

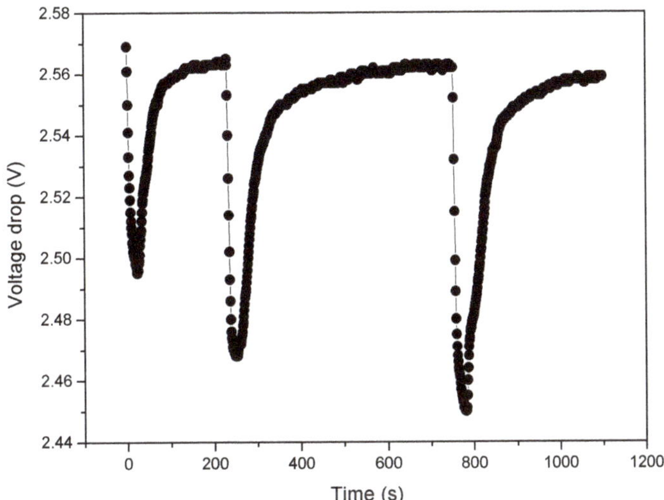

Figure 5. Human-breathing pattern (exhalations) obtained by the voltage drop technique.

The I–V characteristics of the square-wave voltage source are shown in Figure 6a. This curve was obtained by applying to the generator a gradually increasing pure resistive load of a precisely known value (tolerance of resistors: ±1%), and measuring the output voltage and the absorbed current by true-RMS digital voltmeter and digital ammeter, respectively. The generator I–V characteristics clearly show a non-ideal behavior, with an electromotive force (EMF) of ca. 2.57 V and an internal resistance value (ratio between the EMF and I_{max}) of ca. 120 kΩ. In particular, the short-circuit voltage (V = 0 and I = I_{max}) and the open-circuit voltage, that is, the voltage without load (V = V_{max} and I = 0), were obtained by extrapolation, since direct measurement of the short-circuit voltage could damage the generator because of the high current intensity flowing in the circuit (overload protection is not present in such simple type of generators), while the open-circuit voltage is limited by the digital multimeter input impedance (usually 10 MΩ at a frequency of 50–60 Hz). In order to avoid the influence of the digital voltmeter input impedance, the voltage and current values were measured for resistive loads inferior to 10 MΩ. As is visible in Figure 6b, the power curve of the square-wave generator shows a dependence of the output power on the applied resistive load, with a maximum value of ca. 14 µW. Such maximum power is achieved with a pure resistive load of 120 kΩ that corresponds exactly to the internal resistance of the generator.

As indicated above, the exhalation pattern displayed as a voltage-drop sequence (see Figure 5) can be converted to a true-RMS current intensity variation or to a normalized impedance variation by using the determined I–V characteristics of the square-wave generator (see Figure 7a,b). In particular, the following relationship was used to convert the time-resolved voltage-drop measurements to time-resolved current-intensity data:

$$I = I_{sc} \cdot \left(1 - \frac{V}{E}\right), \tag{1}$$

where $I_{sc} = I_{max} = 21.3$ µA is the short-circuit current intensity, and E is the electromotive force. Similarly, time-resolved zeolite-sensor impedance (Z) values can be obtained from the experimental voltage data by the following relationship:

$$Z = \frac{V}{I} = \frac{V}{I_{SC} \cdot \left(1 - \frac{V}{E}\right)} = \frac{V}{\left(I_{SC} - \frac{1}{r} \cdot V\right)}, \tag{2}$$

where r = 120 kΩ is the generator internal resistance. As visible, when the breath pattern is expressed as temporal variation of the impedance, the exhalation signals result in being quite deformed (see Figure 7b).

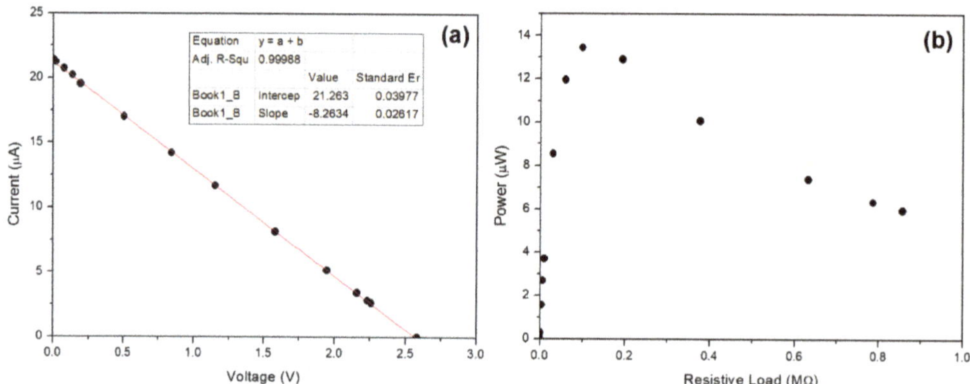

Figure 6. I-V characteristics of the square-wave voltage source (**a**) and its corresponding power curve (**b**).

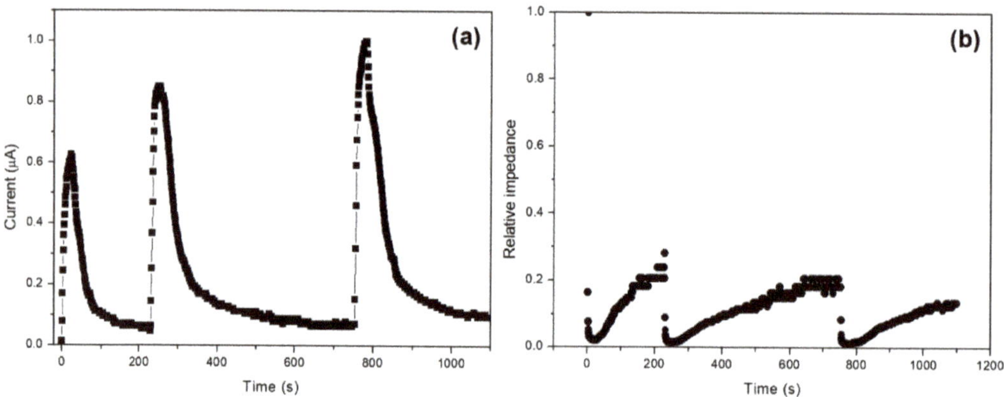

Figure 7. Exhalation signals displayed as time-resolved current intensity (**a**) and impedance (**b**).

The observed very fast response of the clinoptilolite-based sensor to the exhaled humidity is related to the special ionic transport mechanism that involves exclusively extra-framework cations located on the external surface of the zeolite slab. In fact, since the electrical contacts are located on the surface, only surface cations are subjected to the applied alternated electric field. During exhalation, these cations are directly exposed to the external humid environment, and, therefore, they readily interact with water molecular dipole by ion-cation electrostatic forces, thus leading to their hydration (solvation). The hydration process (involving one or more water molecules) significantly reduces the cation charge density, weakening, as a consequence, the Coulomb interaction between cations and the negative charge spread in the closed zeolite framework region (nucleophilic area). The consequent strong increase of cation mobility determines an increase in the current intensity moving on the slab surface, with a drop of the voltage between the silver electrodes. During inhalation, the low humidity content characterizing the environment close to the sensor surface determines a desorption of the water molecules from the cations (ion-dipole is a quite weak physical interaction) with restoration of the original very low surface conductivity of the zeolite slab. Such equilibrium between surface and environmental water molecules allows the electrical monitoring of the moisture fluctuations in the space immediately close to the electrode surface.

4. Conclusions

A variety of ac square-wave voltage sources embedded into commercial entry-level digital multimeters (DMM) that are connected in parallel with a true-RMS voltmeter (voltage-drop method) have been tested to detect the signal coming from a simple clinoptilolite-based impedimetric water-vapor sensor. The most sensible response was observed with a square-wave voltage source of 14 µW (maximum) embedded in the DT830D multimeter model. In particular, the water-vapor sensor was fabricated by cutting a piece of geomorphic natural clinoptilolite stone in form of small slab and painting two 1 mm–spaced rectangular electrodes on its surface. When the sensor was exposed to water vapor contained in the exhaled human breath, the slab surface impedance rapidly decreased, and this event was recorded as a voltage drop by the voltmeter data-logging system. Owing to the clinoptilolite surface hydrophobic nature, during the inhalation stage of the human breathing, water molecules rapidly desorbed from cations located on the external surface of the impedimetric sensor, thus determining a fast increase of its impedance value. The sensor showed a very prompt and completely reversible behavior at room temperature, thus allowing its technological exploitation as breath sensor. For example, this instrumental approach can be conveniently used for human-exhalation detection in apnea-syndrome monitoring. The principal physicochemical characteristics of the geomorphic clinoptilolite used for sensors fabrication were also determined by XRD and SEM/EDS.

Author Contributions: Conceptualization, G.C. and L.N.; Formal analysis, G.C.; Funding acquisition, G.C. Both authors have contributed equally to this research. All authors have read and agreed to the published version of the manuscript.

Funding: This research received no external funding.

Institutional Review Board Statement: Not applicable.

Informed Consent Statement: Not applicable.

Data Availability Statement: Not applicable.

Conflicts of Interest: The authors declare no conflict of interest.

References

1. Park, S.; Choi, K.S.; Lee, D.; Kim, D.; Lim, K.T.; Lee, K.-H.; Seonwoo, H.; Kim, J. Eggshell membrane: Review and impact on engineering. *Biosyst. Eng.* **2016**, *151*, 446–463. [CrossRef]
2. Sajid, M.; Aziz, S.; Kim, G.B.; Kim, S.W.; Jo, J.; Choi, K.H. Bio-compatible organic humidity sensor transferred to arbitrary surfaces fabricated using single-cell-thick onion membrane as both the substrate and sensing layer. *Sci. Rep.* **2016**, *6*, 30065. [CrossRef] [PubMed]
3. Su, B.; Gong, S.; Ma, Z.; Yap, L.W.; Cheng, W. Mimosa-Inspired Design of a Flexible Pressure Sensor with Touch Sensitivity. *Small* **2015**, *11*, 1886–1891. [CrossRef] [PubMed]
4. Potyrailo, R.A.; Ghiradella, H.; Vertiatchikh, A.; Dovidenko, K.; Cournoyer, J.R.; Olson, E. Morpho butterfly wing scales demonstrate highly selective vapour response. *Nat. Photon.* **2007**, *1*, 123–128. [CrossRef]
5. Potyrailo, R.A. Bio-inspired device offers new model for vapor sensing. *SPIE Newsroom* **2011**, *10*, 3568. [CrossRef]
6. Nassar, J.M.; Cordero, M.D.; Kutbee, A.T.; Karimi, M.A.; Torres Sevilla, G.A.; Hussain, A.M.; Shamim, A.; Hussain, M.M. Paper Skin Multisensory Platform for Simultaneous Environmental Monitoring. *Adv. Mater. Technol.* **2016**, *1*, 1600004. [CrossRef]
7. Vukusic, P.; Sambles, J. Photonic structures in biology. *Nature* **2003**, *424*, 852–855. [CrossRef]
8. Hovhannisyan, V.A.; Dong, C.-Y.; Lai, F.-J.; Chang, N.-S.; Chen, S.-J. Natural zeolite for adsorbing and release of functional materials. *J. Biomed. Opt.* **2018**, *23*, 91411. [CrossRef]
9. Ghobarkar, H.; Schäf, O.; Guth, U. Zeolites—From kitchen to space. *Prog. Solid St. Chem.* **1999**, *27*, 29–73. [CrossRef]
10. Sazama, P.; Jirglová, H.; Dedecek, J. Ag-ZSM-5 zeolite as high-temperature water-vapor sensor material. *Mater. Lett.* **2008**, *62*, 4239–4241. [CrossRef]
11. Schäf, O.; Wernert, V.; Ghobarkar, H.; Knauth, P. Microporous Stilbite single crystals for alcohol sensing. *J. Electroceramics* **2006**, *16*, 93–98. [CrossRef]

12. Kasperkowiak, M.; Strzemiecka, B.; Voelkel, A. Characteristics of natural and synthetic molecular sieves and study of their interactions with fragrance compounds. *Physicochem. Probl. Miner. Process* **2016**, *52*, 789–802. [CrossRef]

13. Rhodes, C.J. Properties and applications of Zeolites. *Sci. Prog.* **2010**, *93*, 223–284. [CrossRef] [PubMed]

14. Tekin, R.; Erdogmus, H.; Bac, N. Assessment of zeolites as antimicrobial fragrance carriers. In *Antimicrobial Research: Novel Bioknowledge and Educational Programs*; Mèndez-Villas, A., Ed.; Formatex Research Center: Badajoz, Spain, 2017. [CrossRef]

15. Simon, U.; Franke, M.E. *Ionic Conductivity of Zeolites: From Fundamentals to Applications, Host-Guest-Systems Based on Na-Noporous Crystals*; Laeri, F., Schüth, F., Simon, U., Wark, M., Eds.; Wiley-VCH Verlag GmbH & Co. KGaA: Weinheim, Germany, 2003; pp. 364–378.

16. Tabourier, P.; Carru, J.-C.; Wacrenier, J.-M. Dielectric and far-infrared investigation of cation movements in X-type zeolites. Study of their correlations. *J. Chem. Soc. Faraday Trans. 1 Phys. Chem. Condens. Phases* **1983**, *79*, 779–783. [CrossRef]

17. Freeman, D.C.; Stamires, D.N. Electrical conductivity of synthetic zeolites. *J. Chem. Phys.* **1961**, *35*, 799–806. [CrossRef]

18. Saly, V.; Kocalka, S. Dielectric response of natural clinoptilolite type zeolitic material containing silver iodide. *Chem. Pap.* **1996**, *50*, 328–333.

19. Qiu, P.; Huang, Y.; Secco, R.A.; Balog, P.S. Effect of multi-stage dehydration on electrical conductivity of zeolite A. *Solid State Ion.* **1999**, *118*, 281–285. [CrossRef]

20. Stamires, D.N. Effect of Adsorbed Phases on the Electrical Conductivity of Synthetic Crystalline Zeolites. *J. Chem. Phys.* **1962**, *36*, 3174–3181. [CrossRef]

21. Vučelić, D.; Juranic, N. The effect of sorption on the ionic conductivity of zeolites. *J. Inorg. Nucl. Chem.* **1976**, *38*, 2091–2095. [CrossRef]

22. Urbiztondo, M.; Pellejero, I.; Rodriguez, A.; Pina, M.; Santamaria, J. Zeolite-coated interdigital capacitors for humidity sensing. *Sens. Actuators B Chem.* **2011**, *157*, 450–459. [CrossRef]

23. Carotenuto, G.; Camerlingo, C. Kinetic investigation of water physisorption on natural clinoptilolite at room temperature. *Microporous Mesoporous Mater.* **2020**, *302*, 110238. [CrossRef]

24. Lin, C.C.H.; Dambrowitz, K.A.; Kuznicki, S.M. Evolving applications of zeolite molecular sieves. *Can. J. Chem. Eng.* **2011**, *90*, 207–216. [CrossRef]

25. Carotenuto, G. Electrical Investigation of the Mechanism of Water Adsorption/Desorption by Natural Clinoptilolite Desiccant Used in Food Preservation. *Mater. Proc.* **2020**, *2*, 15. [CrossRef]

26. Kowalczyk, P.; Sprynskyy, M.; Terzyk, A.P.; Lebedynets, M.; Namieśnik, J.; Buszewski, B. Porous structure of natural and modified clinoptilolites. *J. Colloid Interface Sci.* **2006**, *297*, 77–85. [CrossRef] [PubMed]

27. Duan, Z.; Jiang, Y.; Tai, H. Recent advances in humidity sensors for human body related humidity detection. *J. Mater. Chem. C* **2021**, *9*, 14963–14980. [CrossRef]

28. Duan, Z.; Zhao, Q.; Wang, S.; Huang, Q.; Yuan, Z.; Zhang, Y.; Jiang, Y. Halloysite nanotubes: Natural, environmental-friendly and low-cost nanomaterials for high-performance humidity sensor. *Sens. Actuators B Chem.* **2020**, *317*, 128204. [CrossRef]

29. Duan, Z.; Jiang, Y.; Zhao, Q.; Wang, S.; Yuan, Z.; Zhang, Y.; Liu, B.; Tai, H. Facile and low-cost fabrication of a humidity sensor using naturally available sepiolite nanofibers. *Nanotechnology* **2020**, *31*, 355501. [CrossRef] [PubMed]

30. Duan, Z.; Zhao, Q.; Wang, S.; Yuan, Z.; Zhang, Y.; Li, X.; Wu, Y.; Jiang, Y.; Tai, H. Novel application of attapulgite on high performance and low-cost humidity sensors. *Sens. Actuators B Chem.* **2020**, *305*, 127534. [CrossRef]

31. Serban, B.-C.; Cobianu, C.; Buiu, O.; Bumbac, M.; Dumbravescu, N.; Avramescu, V.; Nicolescu, C.M.; Brezeanu, M.; Radulescu, C.; Cracium, G.; et al. Quaternary holey carbon nanohorns/SnO$_2$/ZnO/PVP nano-hybrid as sensing element for resistive-type humid sensor. *Coatings* **2021**, *11*, 1307. [CrossRef]

32. Zhang, Y.; Wu, Y.; Duan, Z.; Liu, B.; Zhao, Q.; Yuan, Z.; Li, S.; Liang, J.; Jiang, Y.; Tai, H. High performance humidity sensor based on 3D mesoporous Co$_3$O$_4$ hollow polyhedron for multifunctional applications. *Appl. Surf. Sci.* **2022**, *585*, 152698. [CrossRef]

33. Matko, V.; Milanovič, M. Detection principles of temperature compensated oscillators with reactance influence on piezoelec-tric resonator. *Sensors* **2020**, *20*, 802. [CrossRef] [PubMed]

34. Mogera, U.; Sagade, A.A.; George, S.J.; Kulkarni, G.U. Ultrafast response humidity sensor using supramolecular nanofiber and its application in monitoring breath humidity and flow. *Sci. Rep.* **2014**, *4*, 4103. [CrossRef] [PubMed]

35. Liao, F.; Zhu, Z.; Yan, Z.; Yao, G.; Huang, Z.; Gao, M.; Pan, T.; Zhang, Y.; Li, Q.; Feng, X.; et al. Ultrafast response flexible breath sensor based on vanadium dioxide. *J. Breath Res.* **2017**, *11*, 36002. [CrossRef] [PubMed]

36. Yang, Y.; Gao, W. Wearable and flexible electronics for continuous molecular monitoring. *Chem. Soc. Rev.* **2019**, *48*, 1465–1491. [CrossRef] [PubMed]

Article

Functional Polymeric Coatings for CsI(Tl) Scintillators

Gianfranco Carotenuto [1], Angela Longo [1], Giuseppe Nenna [2], Ubaldo Coscia [3] and Mariano Palomba [1,*]

[1]   Institute for Polymers, Composites and Biomaterials (IPCB-CNR), National Research Council,
       Piazzale E. Fermi 1, I-80055 Portici, NA, Italy; giancaro@unina.it (G.C.); angela.longo@cnr.it (A.L.)
[2]   Italian National Agency for New Technologies, Energy and Sustainable Economic Development (ENEA),
       Research Centre Portici, Piazzale E. Fermi 1, I-80055 Portici, NA, Italy; giuseppe.nenna@enea.it
[3]   Department of Physics 'Ettore Pancini', University of Naples 'Federico II', Via Cintia,
       I-80126 Napoli, NA, Italy; ubaldo.coscia@unina.it
*   Correspondence: mariano.palomba@cnr.it

Abstract: The handling of inorganic scintillators (e.g., alkali metal halides) can benefit from the availability of polymeric materials able to adhere to their surface. Polymeric materials, such as epoxy resins, can act as protective coatings, as adhesives for photodiodes to be connected with the scintillator surface, and as a matrix for functional fillers to improve the optical properties of scintillators. Here, the optical properties of two epoxy resins (E-30 by Prochima, and Technovit Epox by Heraeus Kulzer) deposited on the surface of a scintillator crystal made of CsI(Tl) were investigated, in order to improve the detection of high-energy radiation. It is found that these resins are capable of adhering to the surface of alkali metal halides. Adhesion, active at the epoxy–CsI(Tl) interface, can be explained on the basis of Coulomb forces acting between the ionic solid surface and an ionic intermediate of synthesis generated during the epoxy setting reaction. Technovit Epox showed higher transparency, and it was also functionalized by embedding white powdered pigments (PTFE or $BaSO_4$) to achieve an optically reflective coating on the scintillator surface.

Keywords: optical-grade epoxy; inorganic scintillator; alkali metal halides; adhesion; interface; Coulomb forces; optical properties

Citation: Carotenuto, G.; Longo, A.; Nenna, G.; Coscia, U.; Palomba, M. Functional Polymeric Coatings for CsI(Tl) Scintillators. *Coatings* **2021**, *11*, 1279. https://doi.org/10.3390/coatings11111279

Academic Editor: Je Moon Yun

Received: 16 September 2021
Accepted: 18 October 2021
Published: 21 October 2021

Publisher's Note: MDPI stays neutral with regard to jurisdictional claims in published maps and institutional affiliations.

1. Introduction

Alkali metal halide crystals, such as NaCl, NaI(Tl), CsI, and CsI(Tl), are excellent optical windows and scintillator materials to detect high-energy radiation (e.g., γ-rays, and X-rays) [1–13]. These inorganic scintillators offer (i) a high output and energy resolution, (ii) a fast and high linear response, and (iii) a very stable light output over a wide range of temperatures; however, they are moisture sensitive and, therefore, quite difficult to handle [4]. In addition, the difficult processing of these materials strongly limits their technological exploitation. Usually, to solve these problems, the crystal is covered by PTFE tape; however, this technological approach has several limitations (tape breaking, non-uniform shape, etc.). Therefore, in this paper, we propose replacing the tape with structural adhesive polymeric materials such as epoxy resins containing reflective powders.

It is known that ionic scintillators are solids made of close-packed alkali metal cations and halogen anions, interacting by Coulomb's electrostatic forces, and adhesion is not a critical issue for ionic solids such as ceramic materials (e.g., potteries and glasses) since their surfaces have a layer of hydroxyl groups that allows a strong grafting with the adhesive phase [14]. However, such a layer is not present on the surface of an alkali metal halide crystal, and therefore the polymer–crystal interfacial adhesion is a very critical issue for these materials.

It is well known that polymers have a more or less effective capability of adhering to different solid surfaces by physical or even chemical interactions, depending on the type of side group [15]. For such a reason, the adhesion of thermoplastic polymers to an ionic surface should improve upon increasing the side group polarity, but a mechanically stable polymer–ionic solid interface can never be achieved by ion–dipole interactions. On the other hand, epoxy resin is an almost universal structural adhesive class, able to guarantee a strong interface with solids such as ceramics, inorganic glasses, metals, and even most plastics [16,17]. Since ionic groups (alkoxide and ammonium) are generated in the epoxy structure during the setting reaction [18], these materials can determine an effective interaction with the cations/anions present on the alkali metal halide surface, which provides the adhesion with the crystal.

Here, the adhesion of two optical-grade epoxy resins to a CsI(Tl) crystal was tested after their characterization by UV–Vis spectroscopy, fluorescence spectroscopy, scanning electron microscopy (SEM), and energy-dispersive X-ray spectroscopy (EDS). To improve the reflectance of the coating layer, these resins were filled by white organic/inorganic pigments of PTFE or $BaSO_4$.

2. Materials and Methods

The experimental activity was focused on the surface modification of inorganic scintillators, based on thallium-doped cesium iodide, CsI(Tl), monocrystals, by epoxy resins. The aim of the surface treatment was the protection/functionalization of the scintillator surface, and it was carried out on single crystals, having a cubic shape with a side of 35 mm. Since CsI is highly hygroscopic and sensitive to oxidation, the scintillators were stored in a desiccator with well-activated silica gel. Two types of optical-grade epoxies: E-30 by Prochima and Technovit Epox by Heraeus Kulzer, were tested as scintillator adhesives. The base resin-to-hardener weight ratio was 5:3 for the E-30 epoxy resin, and 3:1 for Technovit Epox. In the case of Technovit Epox, a fast hardener was used. An as-prepared epoxy resin mixture was applied by spraying it on the crystal surface. After that, to eliminate air bubbles and other defects formed in the deposited layer, the covered crystals were kept in oven, under vacuum, at a temperature of 40 °C (for E-30) and 80 °C (for Technovit Epox), for ca. 2 h, and then the samples were left to cure in air for 2 days. Teflon nanopowder (PTFE, Aldrich, St. Louis, MO, USA) and barium sulphate ($BaSO_4$, 99%, Alfa-Aesar, Haverhill, MA, USA) were selected as a white reflective filler for Technovit Epox. In particular, the filler suspension in the base resin component was treated by an ultrasonic bath for 30 min to improve dispersion. After the hardener addition to the base epoxy, accurate mixture stirring, and its deposition on the CsI(Tl) crystal, the coating was allowed to cure at a temperature of 80 °C for 2 h. To obtain an optimal compromise between visible reflectivity and coating uniformity, it was required for the epoxy-based composites were required to be filled with 5.4% by weight of PTFE or 21.3% by weight of $BaSO_4$.

Microscopic characterization and elemental analysis were performed by a scanning electron microscope (FEI Quanta 200 FEG microscope, Theromo Fischer Scientific, Hillsboro, OR, USA)and energy-dispersive X-ray spectroscopy (Inca Oxford 250, Oxford, UK), respectively. Absorption and emission optical spectroscopy measurements were conducted by a UV–Vis–NIR spectrophotometer (PerkinElmer, Lambda-900, Hong Kong, China), equipped with an integrating sphere (diameter of 15 cm, PerkinElmer, Hong Kong, China), covered by Spectralon, and a spectrofluorometer (PerkinElmer, LS-55, Hong Kong, China), respectively.

3. Results and Discussion

The surface morphology of the "as-received" CsI(Tl) crystals was investigated by scanning electron microscopy (SEM, FEI Quanta 200 FEG microscope, Theromo Fischer Scientific, Hillsboro, OR, USA), and energy-dispersive X-ray spectroscopy (EDS) was used in order to quantify the surface oxidation extent. Figure 1a shows the SEM micrograph of the fractured crystal surface. In the inset, the brittle fracture clearly evidences the many dislocations present in the crystal lattice. The EDS spectrum of the fractured surface is shown in Figure 1b where intensive signals of cesium and iodine elements can be appreciated, while the thallium signal is not visible because it is present in a very small amount. A low-intensity oxygen signal is also present, indicating a slight surface oxidation, and the very small carbon peak is due to the SEM sample preparation.

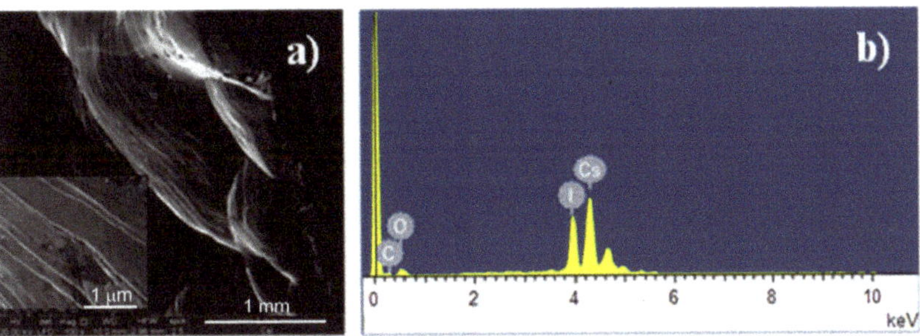

Figure 1. SEM micrograph of the CsI(Tl) crystal surface (**a**) and related EDS spectrum (**b**). The inset of (**a**) shows a magnification of the surface dislocations.

The absorption properties of the CsI(Tl) crystal were measured by absorption spectroscopy. As shown in Figure 2a, even a very thin slice of CsI(Tl) showed a rapidly saturating absorption band starting at 315 nm, while the sample was uniformly transparent in the 320–800 nm spectral range.

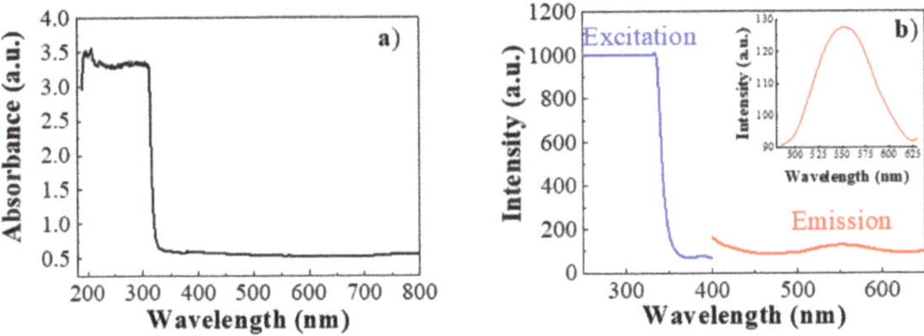

Figure 2. UV–Vis absorption spectrum of CsI(Tl) (slice of 0.883 mm) (**a**), and UV–Vis excitation–emission spectra of CsI(Tl) (**b**). In the inset, a magnification of the emission peak is shown.

The fluorescence characteristics of the scintillator crystals were explored by fluorimetry. As shown in Figure 2b, one visible light emission of green color was found under ultraviolet excitation. In particular, the excitation spectrum (blue curve) shows a cut-off in the UV region, while the emission spectrum (red curve) presents a symmetric peak, centered at 550 nm (see the inset of Figure 2b).

For optimal scintillator operation, it is required to avoid the full attenuation of the fluorescence signal emitted under exposure to high-energy radiation (γ-rays, X-rays, etc.). Therefore, the optical transparency of the epoxy resins (E-30 by Prochima and Technovit Epox by Heraeus Kulzer) in the green spectral range (520–560 nm) was measured. As it can be seen in Figure 3, after curing, these two resins, which were 0.5 mm thick, they had a cut-off value of 272 nm and 290 nm, respectively. Both cut-off values are quite close to the CsI(Tl) crystal self-absorption wavelength, which is 315 nm; however, the optimal optical features were found for the Technovit Epox resin because of the higher transmittance in the UV–Vis–NIR spectral range.

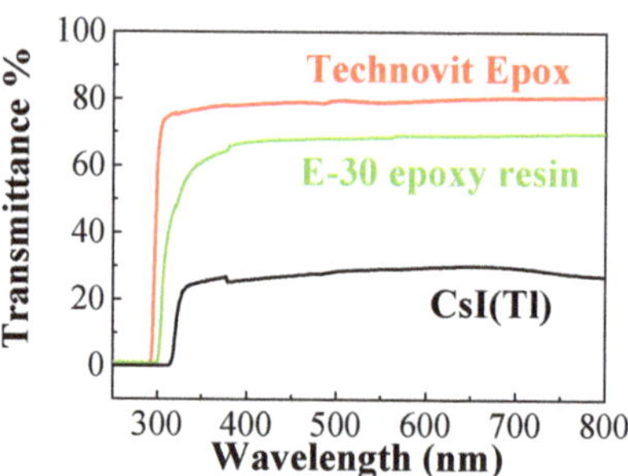

Figure 3. Transmittance spectra of the cured E-30 and Technovit Epox resins, and CsI(Tl) crystal.

Spray technology was used to deposit the Technovit Epox resin on the CsI(Tl) crystal surface. As shown in Figure 4a, a continuous and defect-free coating layer of 0.5 mm thickness was deposited, leading to a coated crystal with a size of 35 mm. Such a type of coating can only slightly attenuate the fluorescent green light emitted by the scintillator crystal under UV radiation (see Figure 4b). According to the difficult peel-out of the Technovit Epox coating, a mechanically robust epoxy–crystal interface resulted. This very good adhesion property could be attributed to some special chemical interaction active at the epoxy–CsI(Tl) interface.

Figure 4. CsI(Tl) scintillator crystal coated by Technovit Epox resin under visible (**a**) and UV light (**b**).

In this work, it was observed that epoxy resin showed good adhesion to the CsI(Tl) crystal surface even without preliminary treatments. Such bonding can be explained on the basis of the formation, during the curing reaction, of an ionic intermediate of synthesis (i.e., -CH(O$^-$)-CH$_2$-NH$_2$$^+$-) that could persist at the organic/inorganic interface because of electrostatic interactions with the Cs$^+$ and I$^-$ ions present on the crystal surface.

To improve the collection of the emitted fluorescence signal by multiple light reflections, the possibility of filling the epoxy layer with a white pigment was also verified. The deposition of such a functionalized coating layer increases the reflectance of the crystal walls.

In the case of the white pigment based on PTFE nanopowder, a not very uniform dispersion was obtained for a filling factor higher than 5.4% by weight, Figure 5a, and large aggregates of PTFE grains appeared in the deposited layer, Figure 5b.

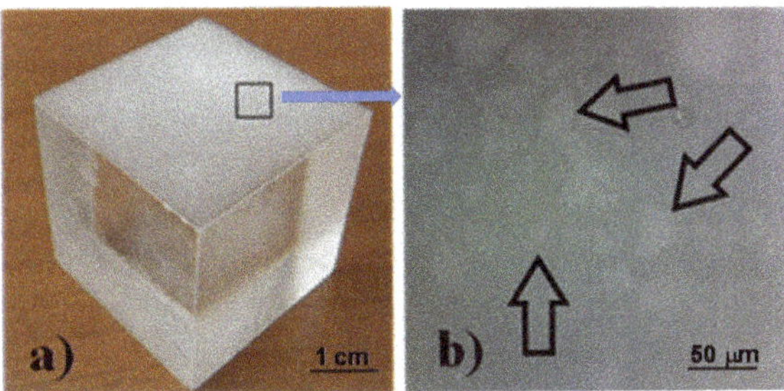

Figure 5. Surface of a CsI(Tl) scintillator coated by a reflective layer of Technovit Epox resin filled with PTFE nanopowder (**a**). An optical micrograph of the achieved surface microstructure, where the arrows indicate the aggregates present in the coating layer (**b**).

To improve the coating uniformity, $BaSO_4$ powder was tested as a filler, since it is a white reflective pigment widely used in optics [19–21]. According to the SEM micrograph and the EDS spectrum shown in Figure 6a,b, the $BaSO_4$ powder had an average size of 431 nm and contained copper impurity (2.3% by weight).

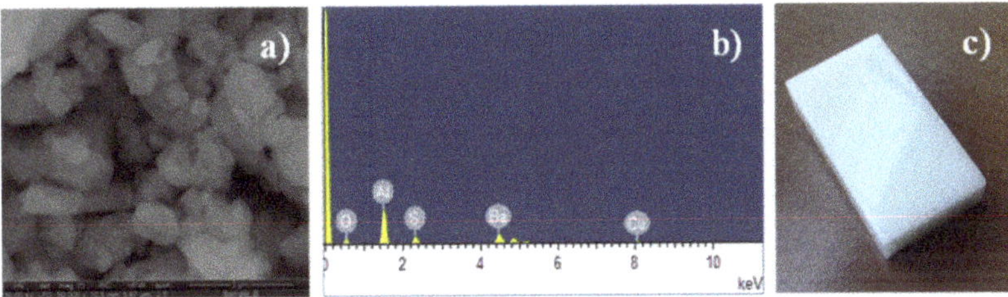

Figure 6. SEM micrograph of the $BaSO_4$ powder (**a**), related EDS spectrum (**b**), and $BaSO_4$/epoxy sample image (**c**).

When sonication was applied during the preparation of the epoxy/$BaSO_4$ mixture, a much higher filling factor was achieved (21.3% by weight), without observing significant grain aggregation in the coating (see Figure 6c).

A reasonably good result was achieved by filling the epoxy resin with $BaSO_4$ powder. Figure 7 shows the total reflectance spectrum in the visible range of the pure CsI(Tl) scintillator crystal (blue curve) and the CsI(Tl) scintillator crystal coated by a $BaSO_4$/epoxy layer (green curve). These optical measurements clearly show a higher reflectance value for the coated crystal.

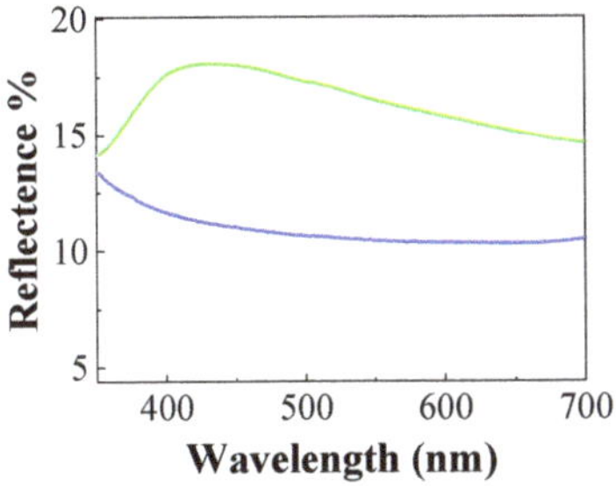

Figure 7. Total reflectance spectra of: CsI(Tl) scintillator (blue curve) and CsI(Tl) scintillator coated by a $BaSO_4$/epoxy layer (green curve).

4. Conclusions

The optical characteristics of two commercial optical-grade epoxy resins used as coatings for CsI(Tl) crystals were compared. The best optical properties were found for Technovit Epox that also showed good adhesion characteristics for the CsI(Tl) crystal surface. Such adhesive properties of epoxies toward alkali metal halides could be ascribed to the possibility for these macromolecules to generate, during the setting reaction, an ionic intermediate that may electrostatically interact with cations and anions at the salt–resin interface. The functionalization the epoxy coating with a white pigment ($BaSO_4$ powder), in order to improve the reflectance on the scintillator surface in the visible spectral region, was also investigated.

Author Contributions: Conceptualization, G.C.; methodology, G.C.; validation, G.C. and U.C.; formal analysis, A.L., M.P. and G.N.; investigation, G.C., A.L., M.P. and G.N.; data curation, G.C., A.L. and M.P.; writing—original draft preparation, G.C., A.L. and M.P.; writing—review and editing, G.C., A.L., M.P., U.C. and G.N. All authors have read and agreed to the published version of the manuscript.

Funding: This research received no external funding.

Institutional Review Board Statement: Not applicable.

Informed Consent Statement: Not applicable.

Data Availability Statement: Not applicable.

Acknowledgments: The authors kindly acknowledge the projects titled: "Nanocompositi polimerici per applicazione ottiche" (CNR) and "Calocube" (INFN). The authors are grateful to Maria Cristina Del Barone of the LAMEST laboratory (IPCB-CNR) for the SEM/EDS measurements.

Conflicts of Interest: The authors declare no conflict of interest.

References

1. Baldochi, S.L.; Ranieri, I.M. *Encyclopedia of Materials: Science and Technology*; Elsevier: Amsterdam, The Netherlands, 2001; pp. 74–78.
2. Phillips, W.M.; Stearns, J.W. Alkali metal/halide thermal energy storage systems performance evaluation. *J. Sol. Energy Eng.* **1987**, *109*, 235–237. [CrossRef]
3. Bourret-Courchesne, E.D.; Bizarri, G.A.; Borade, R.; Gundiah, G.; Samulon, E.C.; Yan, Z.; Derenzo, S.E. Crystal growth and characterization of alkali-earth halide scintillators. *J. Cryst. Growth* **2012**, *352*, 78–83. [CrossRef]
4. Yanagida, T. Inorganic scintillating materials and scintillation detectors. *Proc. Jpn. Acad. Ser. B* **2018**, *94*, 75–97. [CrossRef] [PubMed]
5. Kubota, S.; Sakuragi, S.; Hashimoto, S.; Ruan, J.-Z. A new scintillation material: Pure CsI with 10 ns decay time. *Nucl. Instrum. Meth. Phys. Res. A* **1988**, *268*, 275–277. [CrossRef]
6. Adriani, O.; Albergo, S.; Auditore, L.; Zampa, G.; Basti, A.; Berti, E.; Bigongiari, G.; Bonechi, L.; Bongi, M.; Bonvicini, V.; et al. The CALOCUBE project for a space based cosmic ray experiment: Design, construction, and first performance of a high granularity calorimeter prototype. *J. Instrum.* **2019**, *14*, P11004. [CrossRef]
7. Pacini, L.; Adriani, O.; Agnesi, A.; Zampa, G.; Zampa, N.; Auditore, L.; Basti, A.; Berti, E.; Bigongiari, G.; Bonechi, L.; et al. CaloCube: An innovative homogeneous calorimeter for the next-generation space experiments. *J. Phys. Conf. Ser.* **2017**, *928*, 012013. [CrossRef]
8. Kurman, Y.; Shultzman, A.; Segal, O.; Pick, A.; Kaminer, I. Photonic-crystal scintillators: Molding the flow of light to enhance X-ray and γ-ray detection. *Phys. Rev. Lett.* **2020**, *125*, 040801. [CrossRef] [PubMed]
9. Cabanelas, P.; González, D.; Alvarez-Pol, H.; Boillos, J.M.; Casarejos, E.; Cederkall, J.; Cortina, D.; Feijoo, M.; Galaviz, D.; Galiana, E.; et al. Performance recovery of long CsI(Tl) scintillator crystals with APD-based readout. *Nucl. Instrum. Methods Phys. Res. A* **2020**, *965*, 163845. [CrossRef]
10. Gramuglia, F.; Frasca, S.; Ripiccini, E.; Venialgo, E.; Gâté, V.; Kadiri, H.; Descharmes, N.; Turover, D.; Charbon, E.; Bruschini, C.; et al. Light extraction enhancement techniques for inorganic scintillators. *Crystals* **2021**, *11*, 362. [CrossRef]
11. Li, G.; Lou, J.L.; Ye, Y.L.; Hua, H.; Wang, H.; Han, J.X.; Huang, M.R. Property investigation of the wedge-shaped CsI (Tl) crystals for a charged-particle telescope. *Nucl. Instrum. Methods Phys. Res. A* **2021**, *1013*, 165637. [CrossRef]
12. Knyazev, A.; Park, J.; Golubev, P.; Cederkäll, J.; Alvarez-Pol, H.; Benlliure, J.; Timm, R. Simulations of light collection in long tapered CsI (Tl) scintillators using real crystal surface data and comparisons to measurement. *Nucl. Instrum. Methods Phys. Res. Sect. A* **2021**, *1003*, 165302. [CrossRef]

13. Chen, M.; Sun, L.; Ou, X.; Yang, H.; Liu, X.; Dong, H.; Duan, X. Organic semiconductor single crystals for X-ray imaging. *Adv. Mater.* **2021**, 2104749. [CrossRef] [PubMed]
14. Centelles, X.; Castro, J.R.; Cabeza, L.F. Experimental results of mechanical, adhesive, and laminated connections for laminated glass elements—A review. *Eng. Struct.* **2019**, *180*, 192–204. [CrossRef]
15. Wypych, G. *Handbook of Adhesion Promoters*; ChemTec Publishing: Scarborough, ON, Canada, 2018; Chapter 2; pp. 5–44.
16. Jin, F.-L.; Li, X. Synthesis and application of epoxy resins: A review. *J. Ind. Eng. Chem.* **2015**, *29*, 1–11. [CrossRef]
17. May, C. *Epoxy Resins: Chemistry and Technology*; Routledge: Oxfordshire, UK, 2018.
18. Li, X.S.; Li, M.-S.; Chang, F.-C. Kinetics and curing mechanism of epoxy and boron trifluoride monoethyl amine complex system. *J. Polym. Sci. Part A* **1999**, *37*, 3614–3624. [CrossRef]
19. Mikhailov, M.M.; Yuryev, S.A.; Lapin, A.N.; Lovitskiy, A.A. The effects of heating on $BaSO_4$ powders' diffuse reflectance spectra and radiation stability. *Dye. Pigment.* **2019**, *163*, 420–424. [CrossRef]
20. Mikhailov, M.M.; Yuryev, S.A.; Lapin, A.N. Modelling degradation of optical properties for $BaSO_4$ pigments modified by nanoparticles under irradiation with charged particles. *Nucl. Instrum. Methods Phys. Res. Sect. B* **2019**, *458*, 33–38. [CrossRef]
21. Li, X.; Peoples, J.; Yao, P.; Ruan, X. Ultrawhite $BaSO_4$ paints and films for remarkable daytime subambient radiative cooling. *ACS Appl. Mater. Interfaces* **2021**, *13*, 21733–21739. [CrossRef] [PubMed]

Article

Influence of the Thermomechanical Characteristics of Low-Density Polyethylene Substrates on the Thermoresistive Properties of Graphite Nanoplatelet Coatings

Ubaldo Coscia [1], Angela Longo [2,*], Mariano Palomba [2,*], Andrea Sorrentino [2], Gianni Barucca [3], Antonio Di Bartolomeo [4,5], Francesca Urban [4,5], Giuseppina Ambrosone [1] and Gianfranco Carotenuto [2]

[1] Department of Physics 'Ettore Pancini', University of Naples 'Federico II', Via Cinthia, I-80126 Napoli, Italy; ubaldo.coscia@unina.it (U.C.); giuseppina.ambrosone@unina.it (G.A.)

[2] Institute for Polymers, Composites, and Biomaterials—National Research Council (IPCB—CNR), SS Napoli/Portici, Piazzale Enrico Fermi, 1-80055 Napoli, Italy; andrea.sorrentino@cnr.it (A.S.); giancaro@unina.it (G.C.)

[3] Department SIMAU, Polytechnic University of Marche, Via Brecce Bianche, I-60131 Ancona, Italy; g.barucca@staff.univpm.it

[4] Department of Physics 'E. R. Caianello', University of Salerno, Via Giovanni Paolo II, 132-84084 Salerno, Italy; adibartolomeo@unisa.it (A.D.B.); furban@unisa.it (F.U.)

[5] Superconducting and Other Innovative Materials and Devices Institute-National Research Council (SPIN-CNR), SS Salerno, Via Giovanni Paolo II, 132-84084 Salerno, Italy

* Correspondence: angela.longo@cnr.it (A.L.); mariano.palomba@cnr.it (M.P.)

Citation: Coscia, U.; Longo, A.; Palomba, M.; Sorrentino, A.; Barucca, G.; Di Bartolomeo, A.; Urban, F.; Ambrosone, G.; Carotenuto, G. Influence of the Thermomechanical Characteristics of Low-Density Polyethylene Substrates on the Thermoresistive Properties of Graphite Nanoplatelet Coatings. *Coatings* **2021**, *11*, 332. https://doi.org/10.3390/coatings11030332

Academic Editor: Philipp Vladimirovich Kiryukhantsev-Korneev

Received: 3 February 2021
Accepted: 12 March 2021
Published: 15 March 2021

Abstract: Morphological, structural, and thermoresistive properties of films deposited on low-density polyethylene (LDPE) substrates are investigated for possible application in flexible electronics. Scanning and transmission electron microscopy analyses, and X-ray diffraction measurements show that the films consist of overlapped graphite nanoplatelets (GNP) each composed on average of 41 graphene layers. Differential scanning calorimetry and dynamic-mechanical-thermal analysis indicate that irreversible phase transitions and large variations of mechanical parameters in the polymer substrates can be avoided by limiting the temperature variations between −40 and 40 °C. Electrical measurements performed in such temperature range reveal that the resistance of GNP films on LDPE substrates increases as a function of the temperature, unlike the behavior of graphite-based materials in which the temperature coefficient of resistance is negative. The explanation is given by the strong influence of the thermal expansion properties of the LDPE substrates on the thermoresistive features of GNP coating films. The results show that, narrowing the temperature range from 20 to 40 °C, the GNP on LDPE samples can work as temperature sensors having linear temperature-resistance relationship, while keeping constant the temperature and applying mechanical strains in the 0–4.2 × 10^{-3} range, they can operate as strain gauges with a gauge factor of about 48.

Keywords: graphite nanoplatelet coatings; low-density polyethylene; differential scanning calorimetry; dynamical-mechanical-thermal analyses; thermoresistive properties

1. Introduction

Flexible electronics covers a wide range of applications including solar cells, piezoresistive sensors, strain gauges, displays, health care, industrial automation, robotics, smart textile and others [1–5]. It is a technology for assembling electronic devices on electrically passive substrates. The simplest flexible circuit is formed by a layer of either metal or polymer composite on flexible plastic substrate, such as polyimide, polyester, polyethylene naphthalate, polyetherimide, polydimethylsiloxane along with various fluropolymers and copolymers [1,5].

On the other hand, carbon-based materials, such as carbon black [6], graphite [7], CNTs [8], graphene [9] and reduced graphene oxide [10], play an important role as conductive layers, in preparing effective electrical conductors. Indeed, these materials can be used as fillers also in ceramics and polymers to increase their electrical conductivity [8,11–14]. On other cases, they can be deposited as conductive thin films on polymer substrates by means of various fabrication techniques [15]. Graphene layers and graphene/graphite nanoplatelets (GNP), for example, can be deposited by a top-down approach using the exfoliation of graphite, by mechanical methods [16–19], electrostatic deposition [20], chemical synthesis [7], electrochemical [21] and thermal expansions [22].

The aim of the present work is to explore the morphological, structural, and thermo-resistive properties of graphite nanoplatelet films deposited by a micromechanical technique [19] on low-density polyethylene (LDPE) substrates for possible applications in flexible electronics [23]. LDPE was selected, since it is the most commonly used polymer with applications ranging from film packaging and electrical insulation to containers and piping [24,25]. Moreover, although the properties of this polymer depend on the grade of crystallinity it is a mechanically tough and flexible material. In addition, this selection is corroborate by the good adhesion property between the LDPE surface and coating due to a high surface energy value ~31.9 mJ/m^2 [26], and a strong CH-π hydrogen bond of ca. 1.5–2.5 kcal/mol [27].

It is known that the thermomechanical stability of the substrate is an important technological issue in the application of microelectronics, micro-electromechanical systems, and flexible electronics [28,29]. Therefore, to establish the proper operating temperature ranges of the samples, extensive characterizations of the LDPE were carried out by means of differential scanning calorimetry and dynamic-mechanical-thermal analysis over wide temperature ranges. These thermomechanical characterizations, along with the electrical ones as a function of the temperature, allow us to investigate the influence of the thermal expansion of the polymer substrates on the thermo-resistive behavior of the deposited GNP films.

2. Materials and Methods

The low-density polyethylene (LDPE, Riblene FF20) selected as a substrate was supplied in form of pellets by Versalis (ENI S.p.A., S. Donato Milanese, Italy) with the following properties: density = 0.921 kg/m^3, melt flow index = 0.8 g/10 min (190 °C/2.16 kg) and $T_{melting}$ = 112 °C. The substrates were produced by compression molding using a PV Collin P300 hydraulic hot-press (Dr. Collin GmbH, Maitenbeth, Germany). Prior to use, the LDPE pellets were carefully kept under vacuum for 24 h at 60 °C to remove the adsorbed water. Then, about 3 g of dried pellets were placed between two Polytetrafluoroethylene (PTFE)-coated peel-ply films and sandwiched between two steel plates. Afterwards, they were placed between the two preheated plates (30 × 30 cm^2) of the press and left freely to melt for about 1 min at 180 °C. The plates were brought closer together in order to obtain a force of 10 kN. The pressure was maintained for about 5 min, after which the samples were removed from the press and cooled down to room temperature. It is worth noting that the reduced processing time at 180 °C and the previous drying process carried out at the lowest temperature of 60 °C, ensured that no thermal degradation occurred in the polymeric material allowing to obtain homogeneous films, without bubbles or inclusions.

After quenching, the samples were peeled off from the nonstick peel layer and cut into a circular shape approximately 20 cm in diameter. The LDPE samples were stabilized at room temperature for three days and then measured with a film thickness gauge.

The GNP films were deposited on the LDPE substrates by the micromechanical method [19] summarized as follows. The preparation started from commercial expandable graphite (Asbury Carbon, Asbury, NJ, USA), which was thermally expanded in air at 750 °C for 3 min using a muffle furnace. Such an expansion reaction explosively transformed the graphite flakes into high porous, wormlike structures usually named 'expanded graphite'. A cleaning sonication bath was used to disaggregate the expanded graphite filaments dispersed in a volatile liquid medium (Acetone, Sigma-Aldrich, St. Louis, MO, USA) by means of ultrasounds applied for a few tens of minutes at room temperature.

The resulting powder was dried in air at room temperature to produce a highly porous and fragile graphite aerogel that was mechanically broken to give graphite platelets. In a previous paper [19] we evaluated by TEM and SEM measurements that these platelets can have a high aspect ratio with an average size of ca. 80 μm and thickness ranging from 10 to 200 nm. Then, a colloidal dispersion (33 g/L) was obtained by accurately mixing the dry graphite platelet powder with pure ethanol (100%, Sigma-Aldrich, St. Louis, MO, USA) using a sonication bath; then a soft graphite paste was achieved at room temperature by solvent evaporation.

Samples were obtained by fixing a LDPE substrate with a thickness of 1 mm onto a flat glass and carefully spreading a small amount of paste onto the substrate surfaces using a LDPE sheet as a counterface and applying on it a manual pressure ranging from 3 to 9 kPa. The deposition temperature of the films was ~20 °C.

The adhesion property between the LDPE surface and graphite/ethanol paste combined with the shear stress and friction forces applied during the deposition, allowed to uniformly cover a large area of LDPE substrate with overlapped graphene or graphite nanoplatelets. The thickness of the deposited coating can be roughly controlled by varying the amount of graphite paste and the applied pressure to spread it on the LDPE substrate. The obtained samples were rinsed with acetone to remove residual GNP debris and then dried in air at room temperature.

The surface morphologies of pristine LDPE substrate and GNP coating layer on LDPE substrate were carried out by scanning electron microscopy (SEM), FEI Quanta 200 FEG (Thermo Fisher Scientific, Hillsboro, OR, USA), operating at 10 kV. The samples were mounted on standard aluminum stubs by double coated carbon conductive tabs. In addition, an ultra-thin coating of electrically-conducting metal (gold/palladium) was sputtered on the sample surface to prevent charging of the specimen, which occurs because of the accumulation of static electric fields.

SEM cross section observations were performed by using a Zeiss Supra 40 field-emission electron microscope (Carl Zeiss Microscopy GmbH, Jena, Germany) operating at 7 kV. For SEM cross-sectional analysis, samples were glued between two silicon substrates. The obtained "sandwich" was grinded on the side using abrasive papers with ascending grit in order to achieve a flat and polished surface adapt to investigate the thickness of the GNP film.

High-resolution transmission electron microscopy (HR-TEM) analysis was carried out by using a Philips CM200 microscope (Philips, Amsterdam, The Netherlands) operating at 200 kV and equipped with a LaB_6 filament. For TEM observations, samples were prepared in cross-section by using a Leica EM UC6/FC6 cryo-ultramicrotome.

Samples' structural properties were also investigated by X-ray diffraction (XRD) measurements using a Panalytical, X'PERT PRO diffractometer (Malvern Panalytical, Cambridge, UK) with CuKα radiation ($\lambda = 0.154$ nm) in the range of 2θ from 5° to 60°, scan step = 0.0130°, and full scan time = 18.9 s.

Thermal analysis of pristine LDPE samples was carried out under a nitrogen atmosphere using a differential scanning calorimeter (DSC) Q1000-TA instrument (TA instrument, New Castle, PA, USA). Small LDPE samples (approximately 1–2 mg) were weighed accurately and then placed into DSC crimped aluminum pans ensuring a good sample/pan contact. Non-isothermal cooling-heating-cooling tests were carried out at different scan rates in the temperature range from −110 °C to +180 °C. Before each experiment, the apparatus was calibrated using indium and zinc as reference.

Expansion and contraction behaviors of the LDPE samples were tested under a very small tensile load of 10 mN. Samples with dimensions of $5 \times 20 \times 0.1$ mm^3 were examined in the temperature range from −40 °C to +40 °C approximately, at a rate of 5 °C/min (measurement with TMA 450, TA-Instruments, New Castle, PA, USA).

The dynamical-mechanical analyses of the LDPE samples were performed using a DMA 2980 (TA instrument, New Castle, PA, USA). Samples with dimensions $18 \times 9 \times 0.1$ mm^3 were tested by applying a variable tensile deformation with the frequency of 1 Hz and constant displacement amplitude of 1.0%. The measurements were carried out by using liquid nitrogen as a coolant, to allow scans from $-120\,°C$ to $110\,°C$ at a scanning rate of $3\,°C/min$. In this case a lower scanning rate was preferred to ensure sufficient thermal equilibrium ($\pm0.1\,°C$) between the instrument oven and the sample as well as to highlight the different transitions in the material.

Electrical measurements of GNP on LDPE samples were executed under vacuum (~1 mbar) in a coplanar configuration by silver paint contacts (1 cm long and 1 mm spaced) spread on their surfaces. Current-voltage (I-V) characteristics were taken in a Janis Research ST-500 probe station (Janis Research, Woburn, MA, USA) equipped with four micromanipulators connected to a source-measurement unit (SMU) Keithley 4200-SCS (Tektronix, Inc. Beaverton, OR, USA). The mean resistance of the samples at different temperatures (T) was estimated from each I-V characteristic and monitoring the resistance during a period of 60s in cooling-heating cycles from $-40\,°C$ to $40\,°C$ at the rate of about $5\,°C/min$ in order to compare the electrical and TMA measurements.

3. Results and Discussions

Carbon-based films with good adhesion to LDPE substrates were obtained by means of a micromechanical technique based on the application of shear-stress and friction forces to a graphite platelets/ethanol paste on the surface of the polymeric substrate [19]. SEM-micrograph in Figure 1a shows that the pristine LDPE surface is quite flat with few little blisters and/or wrinkles as defects, generated by the flat plates of the hydraulic hot-press used during the substrate fabrication process. SEM micrograph in Figure 1b reveals that the coating layer deposited on the LDPE is composed of an aggregate of thin platelets (submicrometric thickness). The platelets are irregular in shape and have typical lateral dimensions ranging from about one to a few tens of microns.

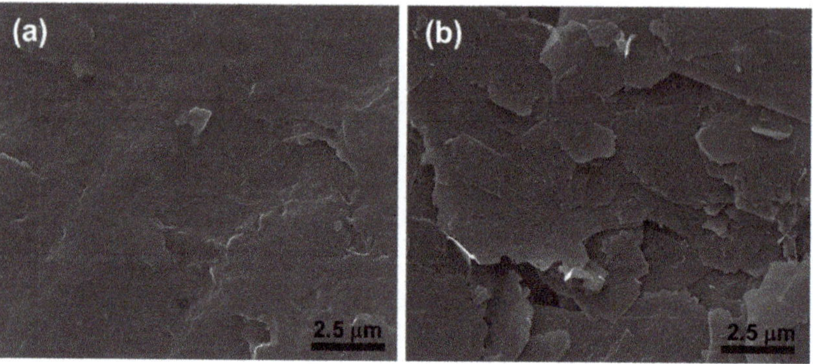

Figure 1. SEM micrographs of low-density polyethylene (LDPE) surface (**a**) and of the coating layer deposited on LDPE (**b**).

In order to investigate the thickness of the coating layer, SEM observations were further performed on cross sectional samples. Figure 2 shows a typical SEM image of such samples. The LDPE substrate is indicated together with the estimated thickness of the coating in three different areas. The sample is slightly tilted with respect to the direction of the electron beam such that it is possible to recognize the shape of some platelets at the top of the coating (red arrows). Some residuals of the glue used to prepare the sample in cross-section (green arrows) are also visible.

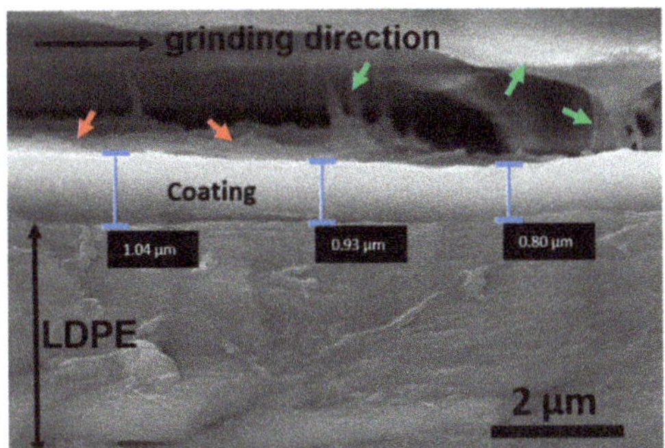

Figure 2. SEM micrograph of a cross-sectional coating deposited on an LDPE substrate. The thickness of the coating is indicated in three different regions.

A major part of the glue was removed by the grind-papers used to obtain a flat surface suitable to reveal the thickness of the coating. In particular, the grinding process was performed in a direction parallel to the LDPE/coating interface, in order to avoid a possible spread of the LDPE substrate in the coating or vice versa. In this way, a selective removal of the softer glue was obtained. SEM measurements allowed to identify the coating layer as a thin film with thickness of (900 ± 200) nm.

The inner structure of the platelets shown in Figure 1b was explored in greater detail by means of HR-TEM measurements performed on cross-sectional samples. HR-TEM analysis reveals that the platelets were composed of stacks of graphene sheets having again a platelet shape and thickness between one and a few tens of nanometers.

According to ref. [30], such nanostructures can be classified as nanoplatelets. Figure 3 shows the two typical configurations of the nanoplatelets inside the larger platelets. They can be identified but not completely separated among them giving rise to a compact structure, as shown in Figure 3a, or they can have a more rarefied nature, as imaged in Figure 3b, where the nanoplatelets are typically thinner and well detached. Inside the nanoplatelets of both the images, fringes corresponding to the different graphene layers can be observed.

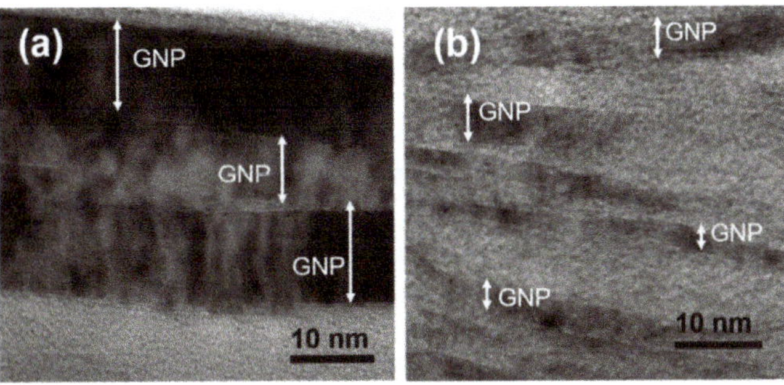

Figure 3. HR-TEM images of cross-sectional film's platelets showing that they are composed by the superposition of graphite nanoplatelets (GNP). GNP can be well compacted (**a**), or quite separated (**b**) inside the platelets.

The structural properties of GNP deposited on LDPE were further investigated by XRD analysis. In Figure 4a the XRD pattern of the pristine LDPE substrate shows a dominant sharp peak at $2\theta = 21.40°$, a weak broad peak at $2\theta = 23.70°$, and one more weak peak at $2\theta = 36.28°$ corresponding respectively to (110), (200), and (020) reflections from the crystalline polyethylene phase [31]. These peaks are superimposed on a large halo, generated by the amorphous polyethylene phase [32,33]. On the other hand, in Figure 4b, the XRD diffractogram of GNP deposited on LDPE includes two peaks of low intensity at $2\theta = 26.46°$ and $2\theta = 42.82°$. These peaks can be referred to the graphite (002) and (100) planes, respectively [19]. As it is known, the average crystal thickness perpendicular to the lattice plane indicated by Miller's indices (hkl) can be obtained from the following Scherrer's equation:

$$L_{(hkl)} = K \times \lambda/(FWHM \times \cos\theta) \tag{1}$$

where FWHM (full-width-at-half-maximum in radians), obtained by a Gaussian fit of the peak, is equal to 0.59 (see inset in Figure 4b), K is a constant depending on the crystallite shape, taken as 0.89 [34,35], λ is the wavelength of the X-ray radiation (Cu-K$_{\alpha 1}$ = 0.15481 nm) and $\theta = 13.23°$ is the half of the corresponding scattering angle. This equation applied to the (002) peak has allowed to determine that the graphite platelets are composed of crystallites with an average size $L_{(hkl)} = 13.6 \pm 0.5$ nm in the (002) direction. Moreover, the interplane distance can be obtained by the Bragg's law for a family of lattice planes (hkl) [35–37]:

$$d_{(hkl)} = n\lambda/\sin\theta \tag{2}$$

where λ and θ are the same parameters previously described, and $n = 1$ is the diffraction peak order.

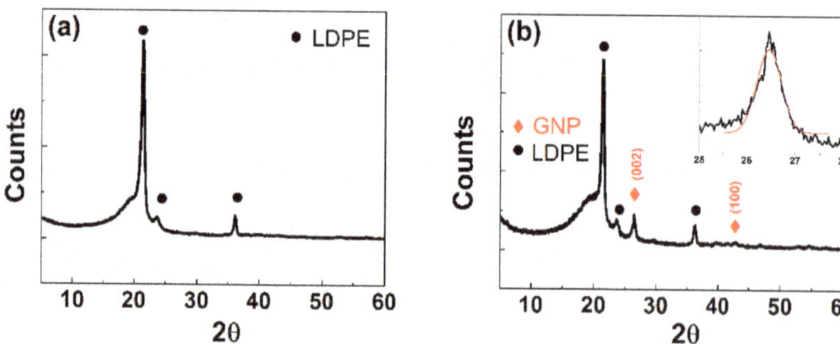

Figure 4. XRD diffractogram of pristine LDPE (**a**), and GNP on LDPE (**b**). The Gaussian fit of (002) peak in the inset of Figure (**b**).

The calculated $d_{(002)}$ value is (0.337 ± 0.003) nm in agreement with the separation distance between two graphene layers ($d = 0.335$ nm) inside the graphite phase [38,39].

Therefore, the average number of the graphene layers (N) per crystalline domain can be calculated from the following equation:

$$N = (L_{(002)}/d_{(002)}) + 1 \tag{3}$$

In our case, the obtained N value is 41, which means, according to the HR-TEM analysis, and following the nomenclature of refs. [30,40] that the deposited film is composed of the superposition of graphite nanoplatelets, each with an average thickness of about 14 nm or, equivalently formed from 41 layers of graphene. All the results are summarized in Table 1. The graphitic nature of the nanoplatelets was confirmed by preliminary Raman measurements carried out on samples prepared in similar conditions.

Table 1. Structural parameters obtained from the (002) peak in the XRD patterns of the GNP on LDPE substrate. 2θ is the diffraction angle, FWHM is the full width at half maximum, $d_{(hkl)}$ is the average distance of graphene layers, $L_{(hkl)}$ is the average height of the crystallites obtained from the FWHM, and N is the average number of graphene layers of a nanoplatelet.

Sample	Miller's Indices (hkl)	Position (2θ)	FWHM (2θ)	$d_{(hkl)}$ (nm)	$L_{(hkl)}$ (nm)	N (counts)
GNP	(002)	26.46	0.59	0.337	13.6	41

In view of their possible applications, to properly define the operating temperature ranges of the samples and avoid irreversible structural modifications of the LDPE capable of causing damage and large fractures also in GNP films, thermal and mechanical properties of the polymer substrates were examined by differential scanning calorimetry (DSC) and dynamic mechanical analysis (DMA) in wide temperature ranges from -110 to $180\,^{\circ}\mathrm{C}$ and from -120 to $110\,^{\circ}\mathrm{C}$, respectively.

Figure 5 show the conventional DSC results from the heating at different scanning rate of the pristine LDPE substrates. The enthalpic curve recorded at $10\,^{\circ}\mathrm{C}/\mathrm{min}$ indicates the occurrence of a broad melting transition with a minimum at about $114\,^{\circ}\mathrm{C}$ (Figure 5a). Probably, the broad melting range is due to the superposition of different melting processes associated with the different crystalline habits formed during the film solidification. Some molecular relaxations with crystals rearrangements are also expected to take place during the melting process. The total heat of fusion associated with this melting peak is $110\,\mathrm{J/g}$, which corresponds to about 37% of the theoretical value found for LDPE sample having 100% of crystallinity ($\Delta H_{\infty} = 293.1\,\mathrm{J/g}$) [41].

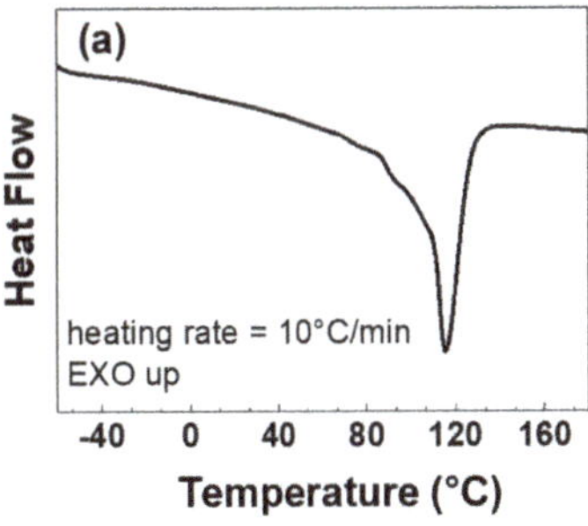

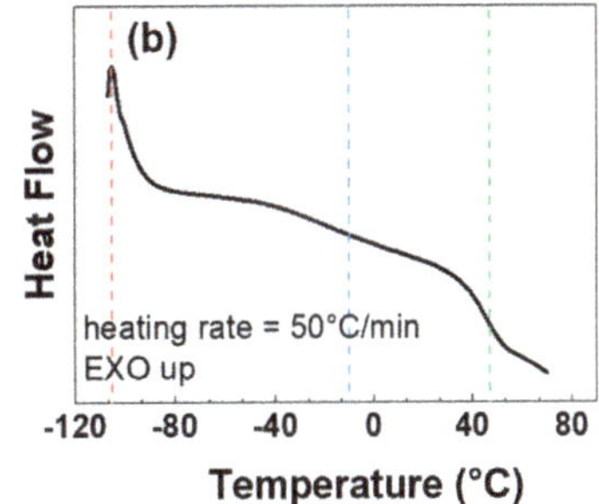

Figure 5. DSC heating curves of a LDPE sample in the -60–$180\,^{\circ}\mathrm{C}$ range at the rate of $10\,^{\circ}\mathrm{C}/\mathrm{min}$ (a) and in the -110–$70\,^{\circ}\mathrm{C}$ range at the rate of $50\,^{\circ}\mathrm{C}/\mathrm{min}$ (b). Vertical dot lines at T = -105, -10 and $47\,^{\circ}\mathrm{C}$ represent the second-order transition temperatures.

The glass transition temperature (T_g) corresponded to the point where half of the increase of the heat capacity occurred in the DSC signal. In order to increase the signal associated with the heat capacity, a DSC scan at high heating rate ($50\,^{\circ}\mathrm{C}/\mathrm{min}$) was carried out (Figure 5b). For LDPE, the enthalpic curve shows three second-order transition temperatures, T = -105, -10 and $47\,^{\circ}\mathrm{C}$ (vertical dot lines in Figure 5b), corresponding to the discontinuity in the second derivative of the Gibbs free energy. These temperatures are associated with different relaxations of the amorphous and crystalline regions present in the polymer matrix. The effects of such transition temperatures on the molecular mobility of the LDPE materials are well evidenced by their thermomechanical behavior.

The dynamic mechanical properties, namely the loss factor as well as the storage modulus and loss modulus as a function of the temperature, are reported in Figure 6a,b, respectively. From the loss factor curve the main features of the LDPE are evident. In particular, the three characteristic relaxation temperatures α, β, and γ of LDPE are clearly resolved. In LDPE, the α relaxation is generally observed between 20 and $60\,^{\circ}\mathrm{C}$ (in this case it corresponds to 47 °C). The intensity of the α relaxation increases with increasing the material density and it is very sensitive to the thermal history of the sample [41,42]. It is associated with the viscoelastic process due to the molecular motion in the crystal caused by the premelting process. The β relaxation generally occurs between −35 and −5 °C and it is related to the amorphous phase. From the loss factor curve shown in Figure 6a the β relaxation for the considered LDPE sample occurs at about −10 °C. Many authors considered β relaxation as the glass transition of the crystalline-amorphous interphase [43,44]. More specifically, it is ascribed to the movement of amorphous molecular chains closer to the crystalline domains. The greater the intensity of this relaxation, the higher is the amorphous fraction present in the system. The γ relaxation for LDPE generally occurs from −95 to −115 °C and it depends on the material density. In this case it occurs at −105 °C. This relaxation is caused by short-range segmental motion. Experimental results obtained on quasi-amorphous samples have demonstrated that the γ transition arises from contributions of the noncrystalline regions of the polymer. This observation favors the assignment of the γ transition rather than the β one as the primary glass transition of LDPE [45]. Despite this discrepancy in the definition of the primary glass transition of linear polyethylene, the phase transitions identified by these temperatures have a strong effect on the mechanical and thermal expansion of LDPE. For example, the thermomechanical behavior of the LDPE (namely the storage modulus, Figure 6b) against the temperature can be divided into distinct regions. These regions are directly related to the mobility of its molecular chains. At very low temperatures, LDPE behaves like a glass and exhibits a high modulus. As increasing the temperature, the modulus continuously decreases. The fall in modulus through the β relaxation region is between one and two orders of magnitude. Upon heating both storage modulus and loss modulus decrease because less force is required for deformation. However, as the temperature approaches the glass transition temperature more work of applied force is dissipated as heat. It results in a local increase in the loss modulus followed by a sharp decrease. The location of the loss modulus peak, which reflects the segmental motion of the LDPE, is considered another physical evidence of the glass transition temperature.

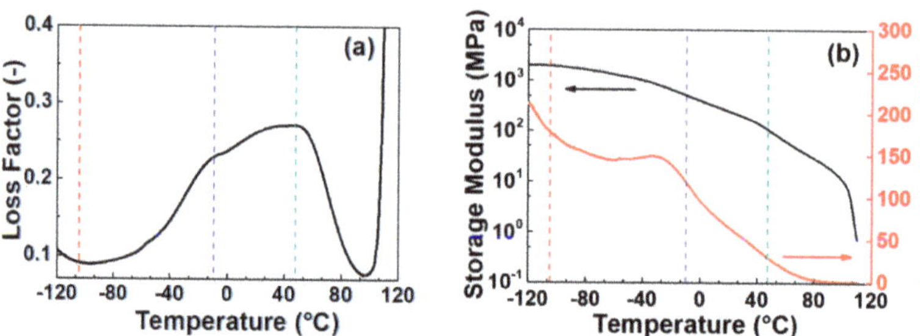

Figure 6. Dynamical mechanical analysis as a function of temperature of a LDPE sample. Loss factor (**a**), storage modulus and loss modulus (MPa) (**b**). Vertical dot lines at T = −105, −10 and 47 °C represent the second-order transition temperatures.

For applications of the LDPE requiring thermal stability and thermomechanical reproducibility, the above study indicates that it is necessary to avoid temperatures that can induce large variations in the mechanical parameters, and irreversible structural changes. To this aim, in the present work, the operating temperature range of the LDPE was roughly limited between −40 and 40 °C to explore the thermal expansion properties. In fact, the melting and glass phase transitions as well as the α relaxation are prevented in this range, while β relaxation (glass transition of the crystalline-amorphous interphase) can take place without modifying the polymer structure. Moreover, a low variation in storage modulus (only about one order of magnitude) occurs and, therefore, suitable working conditions can be obtained in this temperature range for LDPE.

Results on thermal expansion characterization carried out on a typical LDPE substrate sample are presented in Figure 7, where the temperature dependence of the linear strain, ε, is plotted as a function of temperature during a cooling-heating-cooling cycle. Here, ε is calculated by the following ratio:

$$\varepsilon = (l_0 - l)/l_0 \tag{4}$$

where l_0 is the initial size of the sample at temperature $T_0 = 20\ °C$ and l is the size of the sample at the temperature, T. The plot of Figure 7 shows a clear hysteresis between the cooling and heating curves. In particular, the curve recorded in cooling mode is linear, while during the heating the curve shows a continuous change in slope. This is due to the presence of the second-order transition temperature, β, which separates the experimental data into two sets. In fact, as shown in Figure 7, the abscissa of the intersection point of the two linear fit lines obtained in different regions of the heating curve is very close to the β transition temperature ($-10\ °C$). The different trend observed during the heating stage can be explained considering the relaxation time associated with the second order transition. During the cooling stage, the LDPE molecules result undercooled with respect to their equilibrium state. In fact, if the sample is left for a sufficient time at the minimum temperature (in this case $-43\ °C$) it naturally returns to the equilibrium value, corresponding to the initial one of the heating curves. The time requested to reach equilibrium depends on the temperature, the higher temperatures the shorter times are required. For this reason, during heating the polymer molecules have more change to follow point by point their equilibrium state. This behavior is intimately linked to the time-temperature dependence of the molecular mechanisms involved in this second order transition. According to this, the point that identifies the transition represents the temperature at which the kinetics of the process equals the time scale of the experiments (higher heating rate leads to higher temperature transitions). In any case, thermal expansion measurements are always obtained in nonequilibrium conditions, i.e., at finite rates of temperature change. Under these conditions, a correct identification of a transition temperature requires the specification of all the experimental conditions adopted.

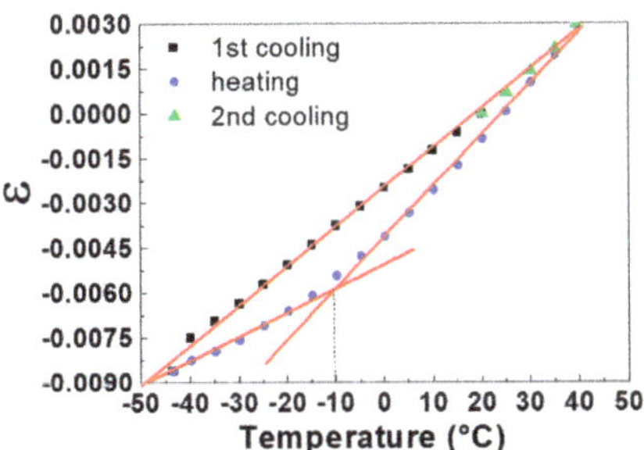

Figure 7. Linear strain, ε, vs. temperature of a LDPE sample subjected to a cooling-heating-cooling cycle. The solid straight lines are linear fit lines, and the vertical dot line allows to determine the β second-order transition temperature.

From the linear fits shown in Figure 7, it is possible to determine, in different temperature ranges, the values of the linear thermal expansion coefficient, CTE, defined as the fractional change in length for a unit change in temperature. The obtained CTE values vary between 1.29×10^{-4} and $1.79 \times 10^{-4}\ °C^{-1}$ according to the literature data [46].

In summary, the characterizations carried out on the LDPE, indicate that in the temperature range between -40 and $40\ °C$ approximately, the constancy in the crystalline phase and the low variation in storage modulus (about one order of magnitude) ensure sufficient stability and reproducibility of the LDPE thermomechanical parameters, despite the occurrence of hysteresis phenomena during the thermal cycles.

Accordingly, the electrical measurements as a function of temperature of GNP films deposited on LDPE substrates were performed from −40 to 40 °C under vacuum in two probe configurations. The ohmicity of the contacts was revealed by the linearity of the I-V characteristics for each temperature value as shown in Figure 8a, where the current I of a representative sample is plotted as a function of voltage V at −40, 20 and 40 °C.

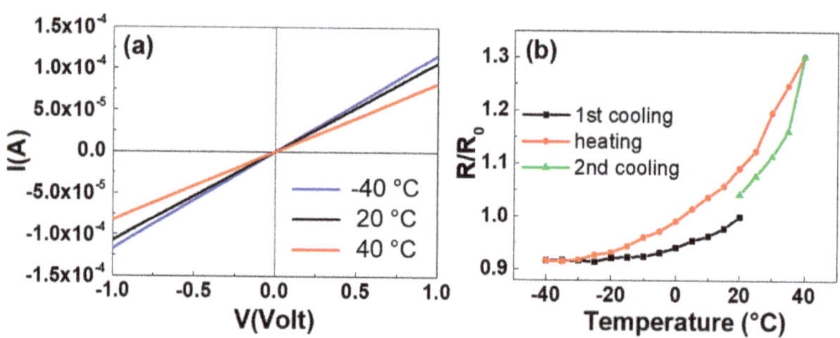

Figure 8. Current-voltage (I-V) characteristics recorded at different temperatures of a GNP film deposited on LDPE substrate (**a**). The ratio of the resistance R to the initial resistance R$_0$ at 20 °C (R/R$_0$) vs. temperature during a cooling-heating-cooling cycle for a GNP film deposited on LDPE substrate (**b**).

The resistances R of the GNP films, determined from the fit of the I-V experimental data plotted in Figure 8a at different temperatures, were $R(20\ °C) = R_0 = 9371 \pm 1\ \Omega$, $R(40\ °C) = 12208 \pm 3\ \Omega$, and $R(-40\ °C) = 8592 \pm 2\ \Omega$. The thermo-resistive properties of GNP films were further explored by recording R as a function of the temperature, starting from 20 °C and performing cooling-heating-cooling cycles in the range between −40 and 40 °C. Figure 8b displays the temperature behavior of the R/R_0 ratio (where R_0 is the initial resistance measured at 20 °C) of the above representative sample under a thermal cycle. Clearly, the resistance increases as a function of the temperature in the investigated range. Such a trend is opposed to the behavior of graphite-based materials, whose resistance decreases with increasing temperature [47]. The resistance increase was observed in all the deposited samples and it can be attributed to the thermal expansion coefficient mismatch between the LDPE ($1–2 \times 10^{-4}\ °C^{-1}$) [46] and nanoplatelet films (approximately $-2 \times 10^{-6}\ °C^{-1}$) [48,49]. Indeed, the larger thermal expansion of the polymeric substrate can induce strains in the deposited film that affect the overlapping GNP and modify the film resistance [50]. Furthermore, it can be seen that the slope of the heating curve between −40 and 20 °C is lower than that between 20 and 40 °C. This behavior can be explained taking into account that the GNP film was deposited at 20 °C, therefore above this value the main effect of the temperature increase is the thermal expansion (larger for LDPE with respect to GNP film) which tends to decrease the contact area among the nanoplatelets and to increase their separation distance. The result is to decrease the number of conductive paths and so to increase the film resistance. When the temperature of the sample is below the deposition temperature (20 °C), the larger contraction of the LDPE compared to that of the GNP film (due again to the difference in their thermal expansion coefficients) leads to a greater compaction of the nanoplatelets. This phenomenon implies that, by heating the sample from −40 to 20 °C, the effect of the resistance decrease with the temperature (characteristic of the graphite/graphene based materials and in any case present) becomes more evident and opposes to the resistance increase induced by the thermal expansion of the substrate. The superimposition of the two opposite resistance behaviors results in a slower growth of total resistance [51].

The resistance hysteresis occurring in GNP film on LDPE during the thermal cycles (Figure 8b) can be attributed to the mismatched thermal expansion coefficient between LDPE and GNP film, too. Indeed, the larger strain of the polymer substrate causes a greater mobility of the nanoplatelets and the occurrence of nano-/microfractures in the films which can lead to different rearrangements of the platelets with different resistance values for the same temperature during a thermal cycle. It was observed that the cooling and heating R-T curves tend to become closer in thermal cycles performed in narrower temperature intervals, however more work needs to be done to reduce the resistance hysteresis in all the examined range by optimizing the deposition conditions of GNP films on LDPE, and to determine the performance of the samples as a function of the film thickness.

It is worth noting that by further limiting the temperature variations between 20 and 40 °C, the fractional change of the electrical resistance, $(R - R_0)/R_0$, is directly proportional to temperature as illustrated in Figure 9a, and the investigated sample works as a temperature sensor. On the other hand, the heating of the sample produces positive tensile strains, ε, which are also directly proportional to temperature, as it can be seen in the same figure.

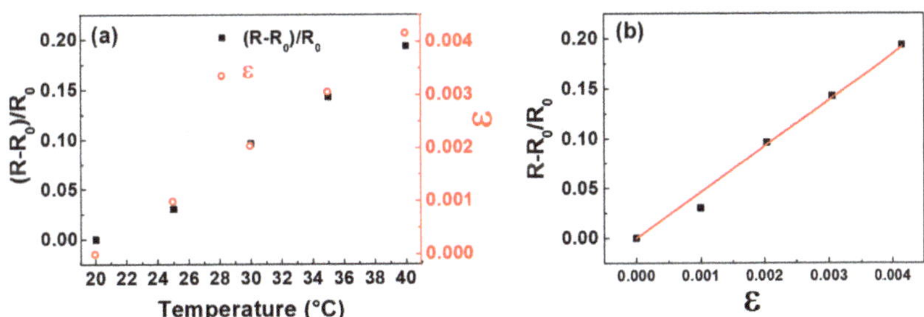

Figure 9. Fractional change of the electrical resistance, $(R - R_0)/R_0$, and strain, ε, versus temperature for a GNP film deposited on LDPE substrate (**a**). Fractional change of the electrical resistance, $(R - R_0)/R_0$, versus strain, ε, for a GNP film deposited on LDPE substrate (**b**). The straight line is the fit of the experimental data.

Therefore, keeping constant the temperature at 20 °C and applying mechanical strain, ε, (having the same values induced by the thermal ones) in the 0–4.2 × 10^{-3} range, $(R - R_0)/R_0$ is a linear function of ε, as shown in Figure 9b. This demonstrates that the sample can operate as a strain gauge because it converts the mechanical strains into electrical resistance changes. The electrical response to mechanical deformations of strain sensors is quantified by the gauge factor (GF), that is the ratio of the relative change in electrical resistance, R, to the mechanical strain, ε, namely

$$GF = (R - R_0)/(\varepsilon R_0) \tag{5}$$

where R_0 is the unstrained resistance. From the fit of the experimental data plotted in Figure 9b, a gauge factor of 48 ± 2 is obtained with a correlation coefficient $r = 0.990$.

4. Conclusions

Films were deposited on LDPE substrates by spreading a graphite paste on their surfaces by using a micromechanical method. SEM, TEM and XRD characterizations show films about 900 nm thick consisting of overlapped nanoplatelets, each composed of 41 graphene layers on average.

DSC and DMA results reveal that by limiting the temperature variations in the range between −40 and 40 °C approximately, irreversible phase transitions, as melting, glass and partial melting/recrystallization, as well as large variations of the mechanical parameters in the LDPE substrates, can be avoided. This prevents damages and large fractures even in the graphite nanoplatelet films. In this temperature range only the glass transition of the crystalline-amorphous interphase (β relaxation) can take place without modifying the polymer structure. However, it leads to a hysteretic behavior in the strain versus temperature curve and variations of the linear thermal expansion coefficient in the range

1.29×10^{-4}–1.79×10^{-4} $^{\circ}C^{-1}$ during thermal cycles. Nevertheless, the constancy in the crystalline phase and the low variation in storage modulus ensure sufficient stability and reproducibility of the LDPE thermo-mechanical parameters.

The electrical measurements of GNP on LDPE samples performed under vacuum in two probe configurations in the temperature range from -40 to 40 $^{\circ}C$, show linear I-V characteristics for each temperature. The electrical resistance of these samples increases with temperature, showing an opposite behavior with respect graphene/graphite-based materials in which the temperature coefficient of resistance is negative. The observed resistance trend can be attributed to the much higher thermal expansion coefficient of the LDPE substrate which, as the temperature increases, induces large strains in the deposited film affecting the overlapping of graphite nanoplatelets and decreasing the number of conductive paths, thus increasing the film resistance. Furthermore, the large strains of the polymer substrate cause a great mobility of the nanoplatelets and the occurrence of nano-/microfractures in the films that can lead to different rearrangements of the platelets with different resistance values for the same temperature during thermal cycles and therefore to hysteresis phenomena.

The hysteresis of the resistance tends to reduce as the temperature range narrows and the fractional change of the electrical resistance becomes directly proportional to temperature between 20 and 40 $^{\circ}C$, indicating that GNP on LDPE samples can work as temperature sensors. Finally, it has been demonstrated that keeping constant the temperature at 20 $^{\circ}C$ and applying mechanical strains in the 0–4.2×10^{-3} range, these samples can operate as strain gauges with a gauge factor of about 48.

Author Contributions: Conceptualization, U.C., G.C., A.S., G.B., and A.D.B.; software, A.L., A.S., G.B., and F.U.; validation, U.C., A.L., M.P., A.S., F.U., G.B., and G.A.; formal analysis, A.L., M.P., U.C., A.S., and G.B.; investigation, G.B., F.U., A.L., A.S., and M.P.; data curation, G.C., A.L., and M.P.; writing—original draft preparation, G.C., M.P., A.L., A.D.B., G.B., A.S., G.A., F.U., and U.C.; writing—review and editing, A.D.B., G.B., G.A., A.S., A.L., and U.C.; supervision, U.C., G.A., G.C., and A.D.B.; All authors have read and agreed to the published version of the manuscript.

Funding: This research received no external funding.

Institutional Review Board Statement: Not applicable.

Informed Consent Statement: Not applicable.

Data Availability Statement: Not applicable.

Acknowledgments: The authors are grateful to M.C. Del Barone of LAMEST laboratory (IPCB-CNR) for SEM measurements. The authors are also grateful to A. Vanzanella and the electronic workshop of INFN Napoli for the fruitful discussions and their support during the assembling of the experimental setup.

Conflicts of Interest: The authors declare no conflict of interest.

References

1. Costa, J.C.; Spina, F.; Lugoda, P.; Garcia-Garcia, L.; Roggen, D.; Münzenrieder, N. *Flex. Sens.—Mater. Appl. Technol.* **2019**, *7*, 35. [CrossRef]
2. Seo, C.U.; Yoon, Y.; Kim, D.H.; Choi, S.Y.; Park, W.K.; Yoo, J.S.; Baek, B.; Kwon, S.B.; Yang, C.M.; Song, Y.H.; et al. Fabrication of polyaniline-carbon nano composite for application in sensitive flexible acid sensor. *J. Ind. Eng. Chem.* **2018**, *64*, 97–101. [CrossRef]
3. Wang, P.; Hu, M.; Wang, H.; Chen, Z.; Feng, Y.; Wang, J.; Ling, W.; Huang, Y. The evolution of flexible electronics: From nature, beyond nature, and to nature. *Adv. Sci.* **2020**, *7*, 2001116. [CrossRef] [PubMed]
4. Hussain, M.M.; El-Atab, N. *Handbook of Flexible and Stretchable Electronics*, 1st ed.; CRC Press: Boca Raton, FL, USA, 2019.
5. Garidiner, F.; Carter, E. *Polymer Electronics—A Flexible Technology*; iSmithers Rapra Publishing: Akron, OH, USA, 2010.
6. Shintake, J.; Piskarev, E.; Jeong, S.H.; Floreano, D. Ultrastretchable strain sensors using carbon black-filled elastomer composites and comparison of capacitive versus resistive sensors. *Adv. Mater. Technol.* **2017**, *3*, 1700284. [CrossRef]
7. Allen, M.J.; Tung, V.C.; Kaner, R.B. Honeycomb carbon: A review of graphene. *Chem. Rev.* **2010**, *110*, 132–145. [CrossRef]
8. Shi, J.; Li, X.; Cheng, H.; Liu, Z.; Zhao, L.; Yang, T.; Dai, Z.; Cheng, Z.; Shi, E.; Yang, L.; et al. Graphene reinforced carbon nanotube networks for wearable strain sensors. *Adv. Funct. Mater.* **2016**, *26*, 2078–2084. [CrossRef]

9. Novoselov, K.S.; Geim, A.K.; Morozov, S.V.; Jiang, D.; Zhang, Y.; Dubonos, S.V.; Grigorieva, I.V.; Firsov, A.A. Electric field effect in atomically thin carbon films. *Science* **2004**, *306*, 666–669. [CrossRef]

10. Longo, A.; Verucchi, R.; Aversa, L.; Tatti, R.; Ambrosio, A.; Orabona, E.; Coscia, U.; Carotenuto, G.; Maddalena, P. Graphene oxide prepared by graphene nano-platelets and reduced by laser treatment. *Nanotechnology* **2017**, *28*, 224002–224008. [CrossRef]

11. Kumar, S.; Blanchet, G.; Hybertsen, M.S.; Murthy, J.; Alam, M. Performance of carbon nanotube-dispersed thin film transistors. *Appl. Phys. Lett.* **2006**, *89*, 143501. [CrossRef]

12. Gorrasi, G.; Bugatti, V.; Milone, C.; Mastronardo, E.; Piperopoulos, E.; Iemmo, L.; Di Bartolomeo, A. Effect of temperature and morphology on the electrical properties of PET/conductive nanofillers composites. *Compos. Part B Eng.* **2018**, *135*, 149–154. [CrossRef]

13. Guadagno, L.; Raimondo, M.; Naddeo, C.; Di Bartolomeo, A.; Lafdi, K. Influence of multiwall carbon nanotubes on morphological and structural changes during UV irradiation of syndiotactic polypropylene films. *J. Polym. Sci. Part B Polym. Phys.* **2012**, *50*, 963–975. [CrossRef]

14. Pinargote, N.W.S.; Smirnov, A.; Peretyagin, N.; Seleznev, A.; Peretyagin, P. Direct ink writing technology (3D printing) of graphene-based ceramic nanocomposites: A review. *Nanomaterials* **2020**, *10*, 1300. [CrossRef] [PubMed]

15. Giovannetti, R.; Rommozzi, E.; Zannotti, M.; D'Amato, C.A. Recent advances in graphene based TiO_2 nanocomposites ($GTiO_2Ns$) for photocatalytic degradation of synthetic dyes. *Catalysts* **2017**, *7*, 305. [CrossRef]

16. Urban, F.; Lupina, G.; Grillo, A.; Martucciello, N.; Di Bartolomeo, A. Contact resistance and mobility in back-gate graphene transistors. *Nano Express* **2020**, *1*, 010001. [CrossRef]

17. Yi, M.; Shen, Z. A review on mechanical exfoliation for the scalable production of graphene. *J. Mater. Chem. A* **2015**, *3*, 11700–11715. [CrossRef]

18. Hernandez, Y.; Nicolosi, V.; Lotya, M.; Blighe, F.M.; Sun, Z.Y.; De, S.; McGovern, I.T.; Holland, B.; Byrne, M.; Gunko, Y.K.; et al. High-yield production of graphene by liquid-phase exfoliation of graphite. *Nat. Nanotechnol.* **2008**, *3*, 563–568. [CrossRef]

19. Coscia, U.; Palomba, M.; Ambrosone, G.; Barucca, G.; Cabibbo, M.; Mengucci, P.; De Asmundis, R.; Carotenuto, G. A new micromechanical approach for the preparation of graphene nanoplatelets deposited on polyethylene. *Nanotechnology* **2017**, *28*, 194001. [CrossRef]

20. Wgbnn Beidaghi, M.; Wang, Z.; Gu, L.; Wang, C. Electrostatic spray deposition of graphene nanoplatelets for highpower thin-film supercapacitor electrodes. *J. Solid State Electrochem.* **2012**, *16*, 3341–3348. [CrossRef]

21. Zhong, Y.L.; Swager, T.M. Enhanced electrochemical expansion of graphite for in situ electrochemical functionalization. *J. Am. Chem. Soc.* **2012**, *134*, 17896–17899. [CrossRef]

22. McAllister, M.J.; Li, J.L.; Adamson, D.H.; Schniepp, H.C.; Abdala, A.A.; Liu, J.; Herrera-Alonso, M.; Milius, D.L.; Car, R.; Prudhomme, R.K.; et al. Single sheet functionalized graphene by oxidation and thermal expansion of graphite. *Chem. Mater.* **2007**, *19*, 4396–4404. [CrossRef]

23. Ren, J.; Wang, C.; Zhang, X.; Carey, T.; Chen, K.; Yin, Y.; Torrisi, F. Environmentally—friendly conductive cotton fabric as flexible strain sensor based on hot press reduced graphene oxide. *Carbon* **2017**, *111*, 622–630. [CrossRef]

24. Zhong, X.; Zhao, X.; Qian, Y.; Zou, Y. Polyethylene plastic production process. *Insight—Mater. Sci.* **2018**, *1*, 1. [CrossRef]

25. Vasile, C.; Pascu, M. *Practical Guide to Polyethylene*; ISmithers Rapra Press: Shrewsbury, UK, 2005.

26. De Oilveira, M.F. Adhesion mechanisms for automotive plastic parts. In Proceedings of the 11th International Conference Coatings, São Paulo, Brazil, 23–25 September 2009.

27. Nishio, M. The CH/p hydrogen bond in chemistry. Conformation, supramolecules, optical resolution and interactions involving carbohydrates. *Phys. Chem. Chem. Phys.* **2011**, *13*, 13873. [CrossRef]

28. Spearing, S.M. Material issues in microelectromechanical systems (MEMS). *Acta Mater.* **2000**, *48*, 179–196. [CrossRef]

29. Gonzalez, M.; Vandevelde, B.; Christiaens, W.; Hsu, Y.Y.; Iker, F.; Bossuyt, F.; Timmermans, P. Thermo-mechanical analysis of flexible and stretchable systems. In Proceedings of the 11th International Thermal, Mechanical and Multi-Physics Simulation, and Experiments in Microelectronics and Microsystems (EuroSimE), Bordeaux, France, 26–28 April 2011; pp. 1–7.

30. Bianco, A.; Cheng, H.-M.; Enoki, T.; Gogotsi, Y.; Hurt, R.; Koratkar, N.; Kyotani, T. All in the graphene family–A recommended nomenclature for two-dimensional carbon materials. *Carbon* **2013**, *65*, 1. [CrossRef]

31. Alexander, L.E. *X-ray Diffraction Methods in Polymer Science*; Krieger Pub.: Huntington, NY, USA, 1979.

32. Singh, B.P.; Saini, P.P.; Gupta, T.; Garg, P.; Kumar, G.; Pande, I.; Pande, S.; Seth, R.K.; Dhawan, S.K.; Mathur, R.B. Designing of multiwalled carbon nanotubes reinforced low density polyethylene nanocomposites for suppression of electromagnetic radiation. *J. Nanopart. Res.* **2011**, *13*, 7065–7074. [CrossRef]

33. Akinci, A. Mechanical and morphological properties of basalt filled polymer matrix composites. *Arch. Mater. Sci. Eng.* **2009**, *35*, 29–32.

34. Sharma, R.; Chadha, N.; Saini, P. Determination of defect density, crystallite size and number of graphene layers in graphene analogues using X-ray diffraction and Raman spectroscopy. *Indian J. Pure Appl. Phys.* **2017**, *55*, 625–629.

35. Buerger, M.J. *X-ray Crystallography*; Wiley: Hoboken, NJ, USA, 1942.

36. Cullity, B.D. *Elements of X-ray Diffraction, Reading*; Addison-Wesley: Boston, MA, USA, 1956.

37. Bacon, G.E. Unit-cell dimensions of graphite. *Acta Cryst.* **1950**, *3*, 137. [CrossRef]

38. Bacon, G.E. The rhombohedral modification of graphite. *Acta Cryst.* **1950**, *3*, 320. [CrossRef]

39. Kazimi, M.R.; Shah, T.; Jamari, S.S.B.; Ahmed, I.; Faizal, C.K.M. Sulfonation of low-density polyethylene and its impact on polymer properties. *Polym. Eng. Sci.* **2014**, *54*, 2522–2530. [CrossRef]

40. De Sousa, D.E.S.; Scuracchio, C.H.; De Oliveira Barra, G.M.; De Almeida, L.A. Expanded graphite as a multifunctional filler for polymer nanocomposites. *Multifunct. Polym. Compos.* **2015**, 245.

41. Stein, R.S. Time-dependence of crystal orientation in crystalline polymers. *Polym. Eng. Sci.* **1968**, *8*, 259–266. [CrossRef]

42. Lam, R.; Geil, P.H. Amorphous linear polyethylene: Annealing effects. *J. Macromol. Sci. Part. B Phys.* **1981**, *20*, 37–58. [CrossRef]

43. Zakin, J.L.; Simha, R.; Hershey, H.C. Low-temperature thermal expansivities of polyethylene, polypropylene, mixtures of polyethylene and polypropylene, and polystyrene. *J. Appl. Polym. Sci.* **1966**, *10*, 1455–1473. [CrossRef]

44. Davis, G.T.; Eby, R.K. Glass transition of polyethylene: Volume relaxation. *J. Appl. Phys.* **1973**, *44*, 4274. [CrossRef]

45. Boyer, R.F. Dependence of mechanical properties on molecular motion in polymers. *Polym. Eng. Sci.* **1968**, *8*, 161–185. [CrossRef]

46. Kissin, Y.V. *Polyethylene: End-Use Properties and Their Physical Meaning*; Hanser Pub Inc.: Liberty Township, OH, USA, 2013.

47. Iwashita, H.; Imagawa, H.; Nishiumi, W. Variation of temperature dependence of electrical resistivity with crystal structure of artificial products. *Carbon* **2013**, *61*, 602–608. [CrossRef]

48. Mag-isa, A.E.; Kim, J.H.; Oh, C.S. Measurements of the in-plane coefficient of thermal expansion of freestanding single-crystal natural graphite. *Mater. Lett.* **2016**, *171*, 312. [CrossRef]

49. Mag-isa, A.E.; Kim, S.M.; Kim, J.H.; Oh, C.S. Variation of thermal expansion coefficient of freestanding multilayer pristine graphene with temperature and number of layers. *Mater. Today Commun.* **2020**, *25*, 101387. [CrossRef]

50. Palomba, M.; Carotenuto, G.; Longo, A.; Sorrentino, A.; Di Bartolomeo, A.; Iemmo, L.; Urban, F.; Giubileo, F.; Barucca, G.; Rovere, M.; et al. Thermoresistive properties of graphite platelet films supported by different substrates. *Materials* **2019**, *12*, 3638. [CrossRef] [PubMed]

51. Tian, M.; Huang, Y.; Wang, W.; Li, R.; Liu, P.; Liu, C.; Zang, Y. Temperature-dependent electrical properties of graphene nanoplatelets film dropped on flexible substrates. *J. Mater. Res.* **2014**, *29*, 1288–1294. [CrossRef]

Article

Green Solid-State Chemical Reduction of Graphene Oxide Supported on a Paper Substrate

Angela Longo, Mariano Palomba * and Gianfranco Carotenuto

Institute for Polymers, Composites, and Biomaterials, National Research Council, UOS Napoli/Portici, Piazzale Enrico Fermi 1, Portici, 80055 Naples, Italy; angela.longo@cnr.it (A.L.); giancaro@unina.it (G.C.)
* Correspondence: mariano.palomba@cnr.it

Received: 22 June 2020; Accepted: 16 July 2020; Published: 17 July 2020

Abstract: The reduction of graphene oxide (GO) thin films deposited on substrates is crucial to achieve a technologically useful supported graphene material. However, the well-known thermal reduction process cannot be used with thermally unstable substrates (e.g., plastics and paper), in addition photo-reduction methods are expensive and only capable of reducing the external surface. Therefore, solid-state chemical reduction techniques could become a convenient approach for the full thickness reduction of the GO layers supported on thermally unstable substrates. Here, a novel experimental procedure for quantitative reduction of GO films on paper by a green and low-cost chemical reductant (L-ascorbic acid, L-aa) is proposed. The possibility to have an effective mass transport of the reductant inside the swelled GO solid (gel-phase) deposit was ensured by spraying a reductant solution on the GO film and allowing it to reflux in a closed microenvironment at 50 °C. The GO conversion degree to reduced graphene oxide (r-GO) was evaluated by Fourier transform infrared spectroscopy (FT-IR) in attenuated total reflectance (ATR) mode and X-ray Diffraction (XRD). In addition, morphology and wettability of GO deposits, before and after reduction, were confirmed by digital USB microscopy, scanning electron microscopy (SEM), and contact angle measurements. According to these structural characterizations, the proposed method allows a bulky reduction of the coating but leaves to a GO layer at the interface, that is essential for a good coating-substrate adhesion and this special characteristic is useful for industrial exploitation of the material.

Keywords: graphene oxide; green chemical reduction; ascorbic acid; reduced graphene oxide

1. Introduction

In recent decades, graphene and its derivatives have attracted great interest because of their excellent mechanical, electrical, thermal, and optical properties [1–5]. On this basis, graphite oxide, first reported over 150 years ago [6], has re-emerged with an intense research interest due to its possible role as a precursor for a cost-effective mass-production of graphene-based materials. Reduced graphene oxide (r-GO) has mechanical, electrical, and optical properties quite similar to graphene because it has a slightly heterogeneous structure made of graphene-like planes, decorated by a few oxygen-containing chemical groups. Owing to the graphene-like properties, r-GO is a convenient material for many technological fields (e.g., catalysis, energy-storage, electronics, etc.) [7,8]. Several GO synthesis methods, which are alternatives to the original Hammer's method [9–12], and new chemical reduction schemes are being proposed [13–27]. The GO conversion to r-GO usually requires some reducing agents or a high temperature [13–18]. Among these reduction strategies, the chemical approach is very promising for bulk production of r-GO at low cost. Unfortunately, most of the widely used chemical reductants such as hydrazine or hydrazine hydrate are highly toxic and/or explosive, which can potentially induce environmental and safety risks [19]. Moreover, the chemical reduction process of GO usually occurs under liquid-phase conditions by dissolving reducing agent and GO in an adequate solvent. This approach commonly determines the presence of some reduction by-products in the final r-GO material. Consequently, an alternative solid-state reduction technique could lead to an environmentally friendly and contaminant-free product [20].

When GO is deposited in the form of a thin film on a substrate, which provides mechanical stability and flexibility to it, during the reduction process, some other limitations related to the presence of this substrate can arise. For example, a limitation of thermal reduction in the presence of substrate is represented by the operating temperature to be applied, which is dependent on the thermal stability of the substrate. On the other hand, in the photo-reduction method, which is also used for supported r-GO production and typically requires laser radiation, the main limitation is that only the sample surface is involved in the process [21–27]. Consequently, the availability of effective soft-chemistry approaches for converting GO supported on substrates to r-GO are strictly required.

Here, a novel method to reduce GO supported on paper to r-GO has been developed. Such a method is based on the use of L-Ascorbic acid (L-aa) as the reductant [28–30]. L-aa is an effective green chemical reductant having a mild activity and non-toxic property, it is largely employed as a reducing agent and it has been also used when GO is converted to r-GO in liquid-phase [31–33]. L-aa is much more environmentally friendly then typical GO reductants like hydrazine and hydrazine hydrate. In addition, a different study revealed that GO reduced by L-aa achieved a C/O ratio (i.e., 12.5) and conductivity values (i.e., 77 S/cm), which are comparable to those produced by hydrazine [34]. Recently, other green reductants such as L-cysteine [35], Glycine [36], and green tea [37] have been study as reductant for GO, however these compounds have demonstrated to be inferior reducing agents. In our process, the solid GO deposit, in a water swelled form, has been reduced by L-aa, taking advantage of a diffusion-based mass-transport mechanism that is possible for the L-aa molecules, in a closed water-refluxing environment. This ecofriendly technique has shown to be very adequate to produce a bulky, uniform and highly reduced GO layer, preserving the paper substrate. Among the possible porous and flexible substrates, paper is one of the most used and low-cost natural polymers [35]. Cellulose is convenient as GO substrate because the hydroxyl groups present on the fiber surface may ensure an adequate interfacial adhesion. In addition, the paper substrate is interesting because it leads to devices that are light in weight, portable, flexible, foldable, and biodegradable which are strongly needed in fields like microfluidics, sensors, etc. [38–43].

The obtained r-GO layers were characterized by X-ray Diffraction (XRD) and Fourier transform infrared spectroscopy in attenuate total reflectance mode (FT-IR/ATR), to have structural information and to establish the reduction degree. In addition, the morphology was investigated by digital USB-microscopy and scanning electron microscopy (SEM), and changes in the GO deposit surface wettability with the reduction treatment was determined by contact angle measurements.

2. Materials and Methods

GO was synthesized by a modified Hummers' method [11,27], that was based on the oxidation of graphite nanoplatelets (GNP) by $KMnO_4$ (≥99%) and KNO_3 (≥99%) solution in absolute H_2SO_4 and after ca. 1 h the reaction was stopped by adding H_2O_2 (30%). The raw reaction product was repeatedly washed by distilled water. All reactants were provided by Sigma-Aldrich. Then, the concentrated aqueous solution of GO was cast onto a paper substrate and this system was allowed to dry in air at room temperature. A Whatman® quantitative filter grade paper, 42 circles, diameter 42.5 mm, was selected as substrate.

The GO supported on paper was placed inside a closed Petri dish (Pyrex), and thermostated at a temperature of 50 °C by using a hot plate. The sample was periodically sprayed by L-ascorbic acid (L-aa, Sigma-Aldrich, St. Louis, MO, USA, 99%) aqueous solution to reduce GO (see Figure 1). Such treatment lasted for ca. 48 h. In these experimental conditions, an effective mass-transport diffusion of the L-aa reductant was possible inside the GO layer swelled by water. To remove all possible reduction by-products, at the end of treatment, the samples were washed by spraying distilled water on them.

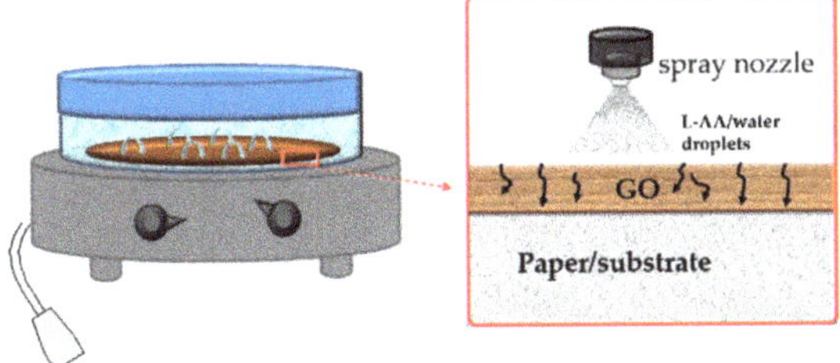

Figure 1. Schematization of experimental set-up.

The surface of samples was analyzed by digital USB-microscope (Dino-lite). In order to establish the time required for a complete reduction, the GO supported on paper samples were characterized by FT-IR spectroscopy, before and after the reduction treatment. FT-IR spectra were recorded in ATR mode in the 4000–700 cm^{-1} range by using a spectrophotometer (PerkinElmer Frontier NIR, Milan, Italy).

The paper substrate and GO supported on paper samples, before and after the reduction treatment, were structurally characterized by XRD (PANalytical-X'PERT PRO diffractometer, with CuKα radiation (λ = 0.154 nm)) in the range of 2θ = 5°–60°, scan step = 0.0130°, and full scan time = 18.9 s.

The surface and the cross-sections morphologies of the coating layer were visualized by a SEM (FEI Quanta 200 FEG, Hillsboro, OR, USA) equipped with an energy dispersive X-Ray (EDS) microanalyzer (Inca Oxford 250, High Wycombe, UK). Cross-sections were obtained by cutting the system with a scalpel.

The surface wettability of GO supported on paper, before and after the reduction treatment, was evaluated by contact angle measurements. In particular, the sessile drop method, that considers the shape of the small liquid test drop to be a truncated sphere, was used. The sessile drop contact angles were measured by a Dataphysics mod. OCA 20, using 1 μL as volume sampling. Each measurement was repeated ten times at room temperature and the achieved contact angle values were statistically treated.

The chemical structures have been drawn by ACD/ChemSketch Freeware Software, version 2019.2.1, Advanced Chemistry Development, Inc., Toronto, ON, Canada, www.acdlabs.com.

3. Results and Discussion

Figure 2 shows the images of a typical sample made of GO supported on paper, before and after the treatment with L-aa aqueous solution. As visible, the GO uniformly covers the surface of the paper substrate and its color changes from a light-brown (see Figure 2a) to black (see Figure 2b), as a consequence of the chemical reduction, thus proving that the GO has been converted to reduced graphene oxide (r-GO). The very uniform black color distribution in the coating layer suggests a complete chemical reduction of the GO into the r-GO. In addition, the coating does not show any defect (e.g., cracks, fractures, pores, etc.) and does not evidence any de-bonding of the substrate, before and after the chemical treatment. Further, the paper substrate was not modified by contacting the acid.

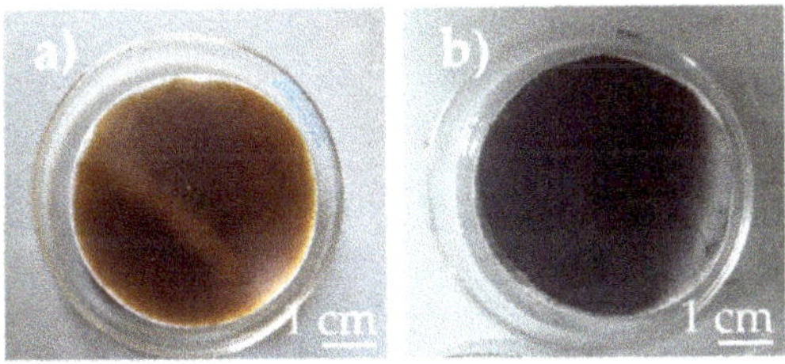

Figure 2. Graphene oxide (GO) supported on paper before (**a**) and after (**b**) treatment with an aqueous L-ascorbic acid (L-aa) solution.

The coating surface morphology, before and after chemical reduction, was investigated by digital USB-microscopy. This type of microscopy is very sensible to the sample reflectivity because the observation takes place under reflected light. As visible in Figure 3, the reflectivity of the two samples significantly differs: the GO sample in Figure 2a does not reflect light and appears mostly brown, differently the r-GO sample in Figure 2b has a much higher reflectivity and includes even silvery areas. It is well known that graphite, similarly to the metals, has a glossy appearance, as a consequence also r-GO resembles graphite and have a mirror-like aspect. The higher reflectivity of the r-GO surface allows to better evidence the presence of an open porosity in the coating layer (dark areas uniformly distributed in the silvery phase).

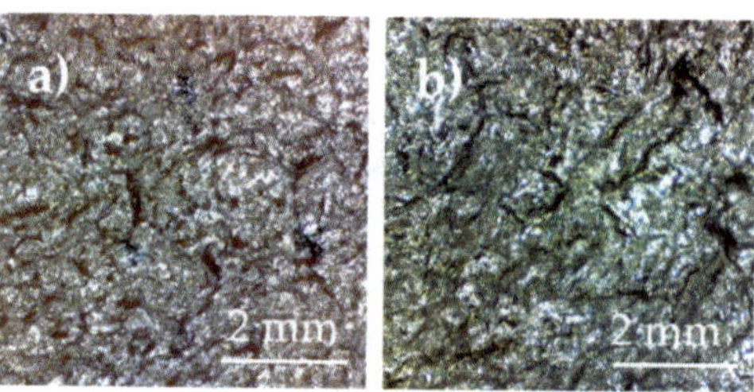

Figure 3. Surface morphologies, obtained by digital USB-microscopy, of: GO coating (**a**) and r-GO coating (**b**) on the paper substrate (100×).

Figure 4 shows the FT-IR spectra in ATR mode of paper substrate (green line), GO before (black line), and after (red line) the treatment with L-aa aqueous solution. The spectrum of paper (green line) shows peaks in the range of 3700–3000 cm^{-1} that can be referred to the resonance vibration of hydroxyl groups (–OH), a peak at ca. 2700 cm^{-1} that is due to the C–H stretching vibration, and other peaks in the range of 1500–900 cm^{-1} which could be due a C–O functional groups. As visible, the GO spectrum includes six main peaks that can be attributed to the resonance vibrations of: (i) the oxygen-containing groups and (ii) the carbon skeleton. In particular, the very strong and broad peak at 3171 cm^{-1} can be referred to the resonance vibration of hydroxyl groups (–OH), the peak at 1735 cm^{-1} to the resonance vibration of carbonyl groups (C=O), the peak at 1227 cm^{-1} to the alloy groups (C–O–H) and the peak at 1061 cm^{-1} to the epoxide groups (C–O–C). In addition, the stretching absorption of C=C groups is also clearly visible in the spectrum at 1621 cm^{-1}, and the band centered at 2840 cm^{-1} is due to the C–H stretching vibration [44].

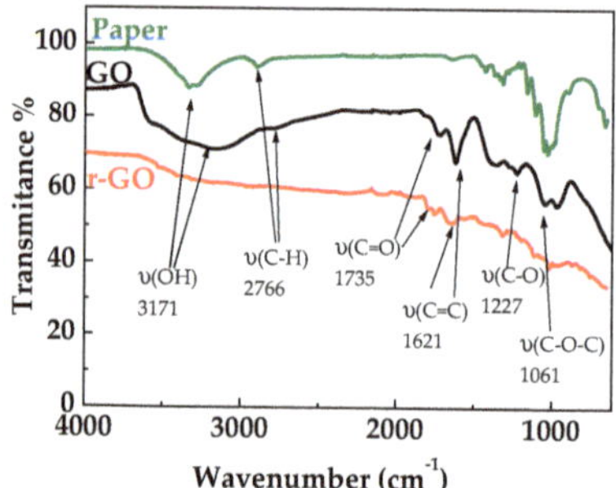

Figure 4. Fourier transform infrared spectroscopy (FT-IR) spectra of paper (green line), GO supported on paper before (black line), and after (red line) the treatment with L-aa aqueous solution.

The spectrum of GO after the chemical reduction (red line in Figure 4) shows a decrease of absorption band number and intensities, thus confirming the formation of r-GO by the removal of most oxygen-containing functional groups. In particular, the C–O stretching resonance completely disappears, and the C=C stretching vibration slightly attenuates and shifts from 1621 to 1590 cm^{-1}, because of the conjugation extension in the graphene plane [44]. The Optimized 3-D images of GO and r-GO are available in the Supplementary Information in Figure S1.

Figure 5 shows XRD diffractograms of: pristine paper substrate, GO, and r-GO supported on paper. The paper substrate has the typical semi-crystalline cellulose diffraction pattern (see Figure 5a), which includes diffraction peaks at 2θ values of: 14.7°, 16.8°, and 22.7°, corresponding respectively to the $(\bar{1}10)$, (110), and (200) crystallographic planes of type I monolithic cellulose (PDF files: 000561717, 000561718, and 000561719) [34]. The XRD patterns of GO and r-GO supported on paper are shown in Figure 5b,c. As visible in Figure 5b, the untreated GO deposit XRD diffractogram includes, in addition to the paper substrate XRD pattern, the presence of a peak at 2θ = 10.62°, referred to the GO (002) plane [14]. The XRD diffractogram of r-GO deposit, shown in Figure 5c, includes, in addition to the paper substrate pattern, a peak of low intensity at 2θ = 11.53°, referred to the GO (002) plane, and two peaks at 25.39° and 43.02°, belonging to r-GO [14]. Other peaks present in the XRD diffractogram can be referred to the presence of inorganic/organic crystalline solid phases (i.e., residual L-aa, calcium salts, etc.) present in the sample at an impurity level. As known, the d-spacing of the GO is higher than the corresponding value of graphite, because of the presence of oxygen-containing functional groups attached on both sheet sides, and of the atomic-scale roughness, arising from structural defects (e.g., sp^3 carbon atoms). It is possible to evaluate this d-spacing by the Bragg's law [45–47]:

$$d_{(hkl)} = n(\lambda/\sin\theta) \tag{1}$$

where λ is the wavelength of the X-ray (Cu-K$_{\alpha 1}$ = 1.5481 Å), θ is half of the corresponding scattering angle, n is an integer number, representing the diffraction peak order, d is the inter-plane distance of the lattice, and (hkl) are the Miller's indices. According to Equation (1), such increased value of the d-spacing, calculated by a GO peak analysis, results: $d_{(002)}$ = 8.32 Å for the untreated GO deposit, while it results 7.67 Å, after the chemical treatment (see Table 1). The calculated d-spacing values were more than twice compared to that of pristine graphite ($d_{(002)}$ = 3.54 Å), thus confirming the expansion of graphitic stack during oxidation. Moreover, Scherrer's equation can be used to obtain the average crystal thickness, i.e., the size perpendicular to the (002) plane (2θ = 10.62° for GO and 2θ = 25.39° for r-GO) and the in-plane crystallite average size that can be expressed in terms of in-plane periodicity peak (2θ = 43.02° for the (100) plane) [30–32] as follows:

$$L_{(hkl)} = K\cdot\lambda/FWHM\cdot\cos\theta \tag{2}$$

where FWHM is the full-width-at-half-maximum in radians obtained by a Gaussian fit of the peak, K is a constant depending on the crystallite shape, and it is taken as 0.89 to evaluate the average in-plane thickness (i.e., $L_{(002)}$), and 1.84, to evaluate the in-plane size (i.e., $L_{(100)}$) [47].

Table 1. Summary of XRD calculations for the system before and after the chemical treatment.

	(hkl)	Position (2θ)	Area (Counts)	FWHM (2θ)	d (nm)	$L_{(hkl)}$ (nm)	n (Counts)
GO	(002)	10.62	2699.59	0.66	0.832	11.72	15
GO	(002)	11.53	876.68	0.39	0.767	19.95	27
r-GO	(002)	25.39	123.05	0.28	0.354	27.21	79
	(100)	43.02	54.36	0.37	-	41.03	-

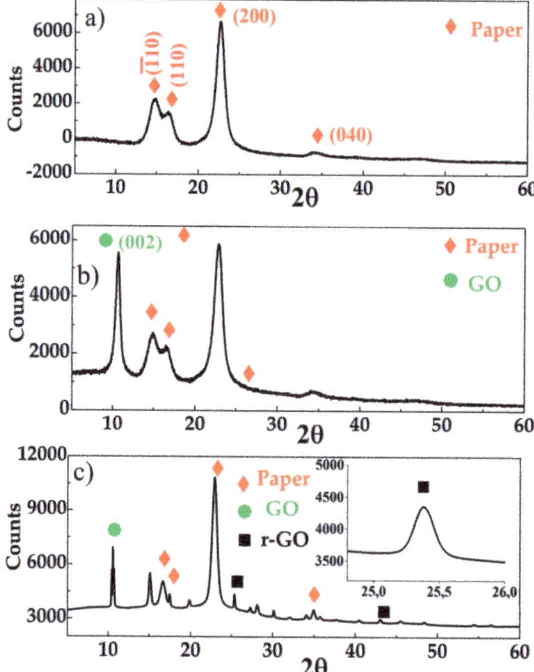

Figure 5. X-ray Diffraction (XRD) diffractogram of the paper substrate (**a**) and GO supported on paper before (**b**) and after (**c**) the treatment with L-aa aqueous solution. The inset in Figure 4c represents a magnification of the r-GO peak (100).

The average number of the graphene layers (*n*) per crystalline domain can be calculate from the XRD peak broadening by using a combination of Scherrer's and Bragg's equations, that provides the following expression:

$$n = (L_{(002)}/d) + 1 \tag{3}$$

Table 1 summarizes the results of these calculations for the system before and after chemical treatment. The (002) peak area of the GO is significantly reduced (68% of the initial area), the average thickness of the GO crystallites before chemical reduction resulted 11.72 nm, while the same thickness was 19.95 nm after the chemical treatment. A thickness value of 27.21 nm was found by analyzing the r-GO peak at $2\theta = 25.39$.

The morphologies of the paper substrate and the GO supported on paper, before and after the chemical reduction, were investigated by SEM, as shown in Figure 6. As visible in the two micrographs showing the substrate microstructure at different magnifications (Figure 6a,b), the substrate was made of compacted cellulose microfibers having a diameter in the 20–50 μm range. Several pores, with a size of ca. 2.5 μm, were uniformly distributed on the substrate surface. After coating the paper substrate by GO and drying this system in air, according to the SEM-micrographs shown in Figure 6c,d, a continuous and uniform coverage of both the cellulose fibers and pores by a thin GO layer was obtained. Obviously, the observed roughness in the GO surface has to be ascribed to the substrate fibrous morphology that is perfectly reproduced because of the conformational flexibility of GO chemical structure for the presence of the sp³-hybridized carbon atoms. Owing to the chemical affinity between the GO layer and the cellulose fibers, related to the presence of hydroxyl groups in both solid phases, a strong interfacial adhesion (i.e., absence of peeling off in the GO coating layer), with complete filling of the pores, was possible. As visible in Figure 6e,f, the coating morphology slightly changed after the chemical reduction treatment, indeed the surface roughness reduced significantly, and all deep grooves and folds are not visible anymore because of the increased stiffness of the graphene coating layer, which consist manly sp^2-hybridized carbon atoms. Such a lower roughness of the sample surface that follows to the chemical treatment proves a complete conversion of GO sheets to r-GO, near the coating top.

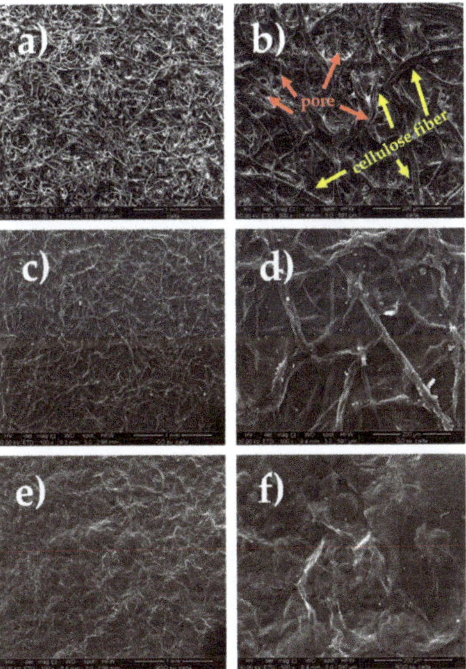

Figure 6. Scanning electron microscopy (SEM) micrographs of paper substrate (**a**,**b**), GO supported on paper (**c**,**d**), and GO supported on paper after treatment with L-aa aqueous solution (**e**,**f**).

A further SEM investigation was carried out on the cross-section of the GO and r-GO supported on paper samples (see Figure 7). According to these images, the GO and r-GO solid phases do not permeate the substrate fibrous structure. Figure 7a,b shows that the thickness of the paper substrate was ca. 120 μm, while the average thickness of GO layer was ca. 3 μm. As visible in Figure 6b, the coating layer appeared as quite compact and cavities or other defects were not present in this layer. In addition, a good interfacial adhesion resulted between the GO substrate and the fibrous coating. The SEM micrographs of the r-GO sample shown in Figure 7c,d suggests the formation of air cavities inside the layer as a consequence of the reduction treatment, and the layer thickness became ca. 50 μm. Probably, such cavities have been produced by the formation of gaseous by-products (mostly H_2O) during the reduction process.

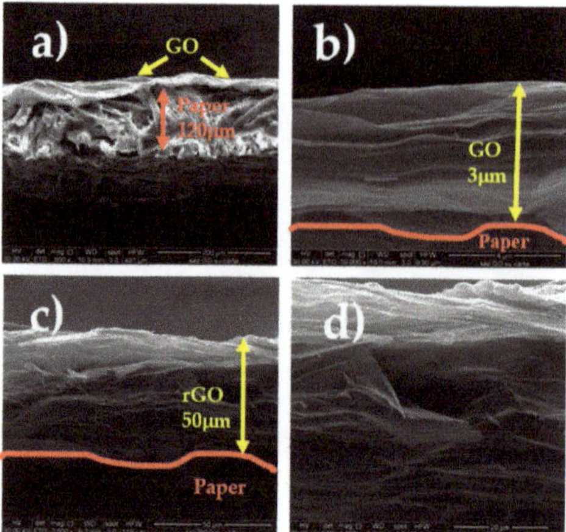

Figure 7. Cross-section SEM-micrographs of GO (**a,b**) and r-GO (**c,d**) supported on paper.

In order to establish the coating layer purity an EDS analysis of the sample, before and after the chemical reduction treatment, has been performed. The EDS spectrum of both GO and r-GO sample is shown in Figure 8. According to this EDS characterization, sulfur impurities (0.75%) and a little amount of silicon (0.20%) were present in the GO coating. However, such impurities completely disappear after the reduction treatment. Owing to the presence of a cellulose substrate, C and O signals remain mostly unchanged by the reduction treatment.

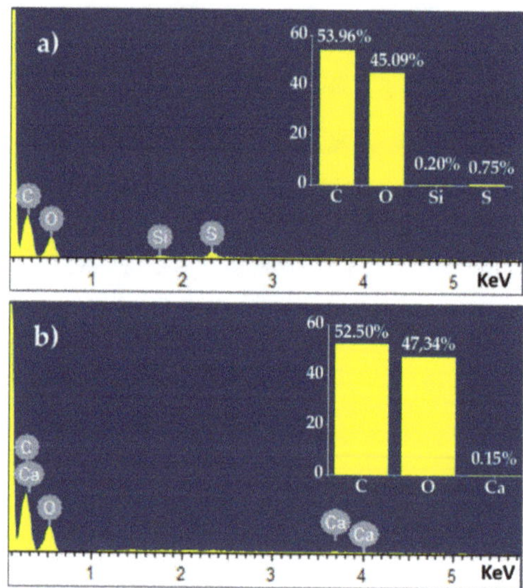

Figure 8. Energy dispersive X-ray spectra of the GO coating layer on paper substrate before (**a**) and after (**b**) the chemical reduction treatment by L-aa (insets give the elemental compositions of the layer).

The variation of the GO coating wettability, as a consequence of the chemical reduction, was quantified by measuring the contact angle of this layer before and after treatment. The "sessile drop method" was used according to the literature [48]; this method considers the shape of the small water test drop as a truncated sphere and measures the angle between the surface of the liquid and the outline of contact surface by image analysis. An example the achieved values of the contact angle are given in

Figure 9. The average contact angle value for a pristine GO layer is 53° ± 1.1°, however, it is possible to observe that the values on the two drop sides are not exactly equal, for example in Figure 9a, on the left side, the average value was 52.6°, while on the right side it was 55.8° and such a discrepancy is related to the high roughness that characterizes the layer surface. This low value of the contact angle confirms the high wettability (hydrophilic nature) of the GO coating. Differently, the average contact angle of a water drop located on the r-GO layer shows higher average values and precisely ca. 97° ± 1.5° (see Figure 9b), thus confirming the low wettability (hydrophobic nature) of the same layer after chemical reduction. This significant increase in the contact angle as a consequence of the chemical treatment shows the efficiency of the proposed method in the removal of most oxygen-containing groups from the GO surface [49].

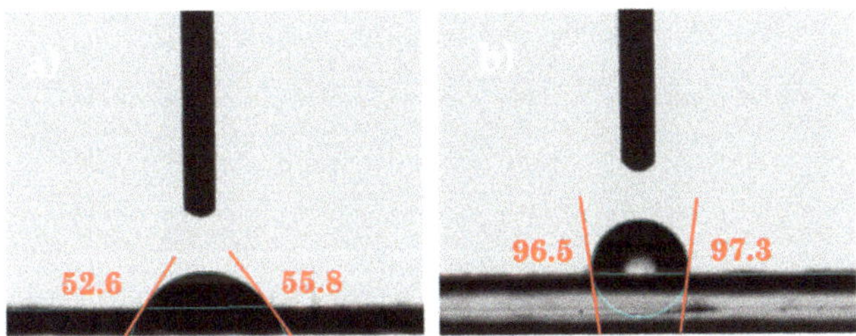

Figure 9. Water contact angle for GO supported on paper before (**a**) and after (**b**) the chemical reduction treatment.

Finally, this chemical technique is based on the possibility to convert the solid GO phase in an aqueous gel, where molecular transport by diffusion easily takes place exposing this hydrophilic substance to water vapor [50]. In particular, the hydrophilic surface of the GO sheets is easily solvated by water molecules that also accumulate at inter-layer, thus generating a continuous liquid phase interpenetrating the GO layers. The use of a temperature close to 50 °C allows a fast diffusion of the L-aa reductant molecules, periodically distributed by spraying a L-aa aqueous solution on the GO layer surface, inside the gel-phase. The mild temperature, selected for the experimental process, allowed to favor the L-aa molecule migration inside the gel-phase, without causing an early thermal degradation of the reductant. The schematic representation of the involved diffusion path is summarized in Figure 10. This process also allowed a solid-state removal of different by-products (e.g., dehydroascorbic acid) by spraying water on the produced r-GO layer.

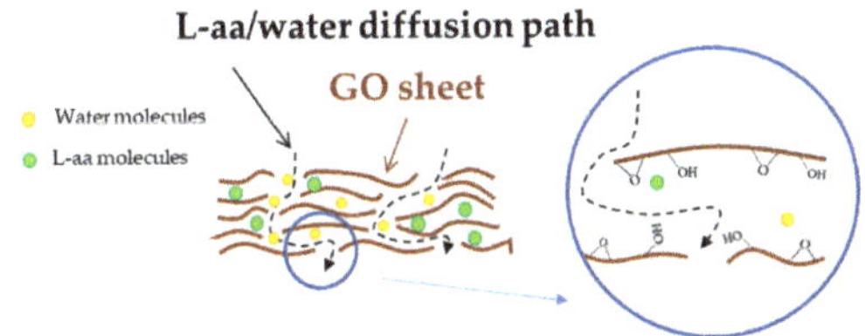

Figure 10. Schematic representation of the L-aa diffusion process inside the swollen GO layer.

Concerning the chemical mechanism taking place in the gel-phase during GO reduction, the investigated system should approximately behave as reported in the literature for the reduction of single GO units dispersed in liquid-phase [51,52]. Owing to the slow diffusion rate of L-aa molecules, which always characterizes a gel-phase reduction, a lower value for the reaction rate is expected. Actually, the exact mechanism of the chemical reaction between GO and L-aa is not completely clear, but two different reactions should be involved in the interaction between L-aa and GO epoxy groups or L-aa and GO vicinal-hydroxyl groups. The reaction pathway is schematically indicated in Figure 11 (3-D images were not energetically optimized). As visible, both reactions require the formation of a good leaving group, which is a hydroxyl group, in the case of epoxies, and a water molecule, in the case of vicinal-hydroxyl groups. These types of leaving group are generated by protonation of the cited GO reactive groups. In particular, the electron withdrawing from five-membered ring of L-aa makes its hydroxyls more acidic, consequently the L-aa is ready to dissociate two protons, that are transferred to GO, thus generating the nucleophilic species (the oxyanion of L-aa, $C_6H_7O_6^-$). After this preliminary acid–base reaction, an SN2 step follows. In this SN2 step, the nucleophilic agent attacks the sp^2-carbon of the epoxy group or the Sp^3-carbon of the alcoholic groups, and hydroxyl or water results as by-product, respectively. In the case of the reaction involving epoxy, a further condensation by SN2 mechanism follows. Then, this intermediate undergoes a thermally induced redox reaction, that leads to formation of reduced graphene (r-GO) and dehydroascorbic acid [51,52]. Optimized 3-D images of different reaction intermediate and some more structural information about the kinetic mechanism are available in the Supplementary Information in Figure S2.

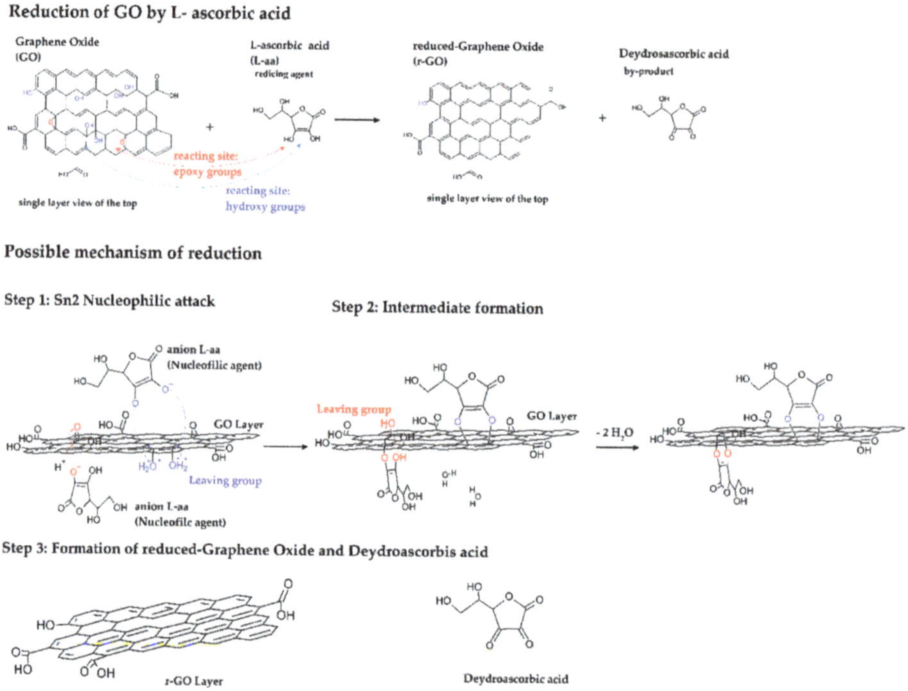

Figure 11. Schematic representation of reaction pathway.

4. Conclusions

The possibility to reduce thin GO deposits on paper substrate by a non-toxic, eco-friendly, and low-cost chemical reductant, like L-aa aqueous solution, has been investigated and according to the achieved results, a uniform and bulky layer of highly reduced r-GO can be obtained. This process requires to reflux the L-aa aqueous solution in presence of the GO deposit in a closed microenvironment for 48 h at a temperature of ca. 50 °C. In these experimental conditions, GO forms a gel-phase where the reductant diffusion can easily take place. According to the SEM investigation, a structural modification of coating surface follows to the chemical reduction treatment, mainly consisting in an increase of the coating flatness. According to the infrared spectroscopy (FT-IR) most of the oxygen-containing groups present in the GO layer were removed, thus determining a significant variation of the wettability of the material, that changes from hydrophilic (contact angle: 53°) to hydrophobic (contact angle: 97°). The XRD results show the presence of the main peaks of the r-GO pattern combined with signals of residual GO, thus confirming the spectroscopic results concerning r-GO formation in presence of a very thin residual GO layer at the interface with paper. In particular, the obtained r-GO coating is composed of graphite platelets, with the average thickness of ca. 27 nm and a width of ca. 40 nm, which are aligned parallel to the interfacial plane and show good graphitic quality. This structural characteristic is relevant to achieve a high flexible r-GO layer, supported on paper substrate because of the excellent GO-paper interface adhesion, which is strictly necessary for an industrial exploitation of this system.

Supplementary Materials: The following are available online at http://www.mdpi.com/2079-6412/10/7/693/s1, Figure S1: Optimized 3-D images of GO and r-GO. Figure S2: Optimized 3-D images of different reaction intermediate and some more structural information about the kinetic mechanism.

Author Contributions: Conceptualization, G.C.; methodology, G.C.; software, A.L.; validation, G.C., M.P., and A.L.; investigation, G.C., M.P., and A.L.; data curation, A.L. and M.P.; writing—original draft preparation, A.L. and M.P.; writing—review and editing, G.C., A.L., and M.P.; supervision, G.C.; project administration, G.C. All authors have read and agreed to the published version of the manuscript.

Funding: This research received no external funding.

Acknowledgments: The authors are grateful to Maria Cristina Del Barone of LAMEST laboratory (IPCB-CNR) for SEM analysis and to Maria Rosaria Marcedula of Thermo-Analysis Laboratory (IPCB-CNR) for FT-IR and contact-angle measurements. The authors acknowledge the funding support by the CNR-project titled: "Nanocompositi polimerici per applicazione ottiche".

Conflicts of Interest: The authors declare no conflict of interest.

References

1. Shafraniuk, S. *Graphene: Fundamentals, Devices, and Applications*, 1st ed.; Jenny Stanford Publishing: Boka Raton, FL, USA, 2015; ISBN 9789814613477.
2. Novoselov, K.S.; Geim, A.K.; Morozov, S.V.; Jiang, D.; Zhang, Y.S.; Dubonos, V.; Grigorieva, I.V.; Firsov, A.A. Electric field effect in atomically thin carbon films. *Science* **2004**, *306*, 666–669. [CrossRef]
3. Geim, A.K. Graphene: Status and prospects. *Science* **2009**, *324*, 1530–1534. [CrossRef] [PubMed]
4. Geim, A.K.; Novoselov, K.S. The rise of grapheme. *Nat. Mater.* **2007**, *6*, 183–191. [CrossRef] [PubMed]
5. Novoselov, K.S.; Jiang, Z.; Zhang, Y.; Morozov, S.V.; Stormer, H.L.; Zeitler, U.; Maan, J.C.; Boebinger, G.S.; Kim, P.; Geim, A.K. Room-temperature quantum hall effect in graphene. *Science* **2007**, *315*, 1379. [CrossRef] [PubMed]
6. Schafhaeutl, C. On the combination of carbon with silicon and iron, and other metals, forming the different species of cast iron, steel, and malleable iron. *Phil. Mag.* **1840**, *16*, 570–590.
7. Smith, A.T.; LaChance, A.M.; Zeng, S.; Liu, B.; Sun, L. Synthesis, properties, and applications of graphene oxide/reduced grapheneoxide and their nanocomposites. *Nano Mater. Sci.* **2019**, *1*, 31–47. [CrossRef]
8. Roweley-Neale, S.J.; Randviir, E.P.; Abo Dena, A.S.; BanKs, C.E. An overview of recent applications of reduced graphene oxide as a basis of electroanalytical sensing platforms. *Appl. Mater. Today* **2018**, *10*, 218–226. [CrossRef]
9. Hummers, W.S.; Offeman, R.E. Preparation of graphitic oxide. *J. Am. Chem. Soc.* **1958**, *80*, 1339. [CrossRef]
10. Zhao, J.; Liu, L.; Li, F. *Graphene Oxide: Physics and Application*; Springer: Berlin/Heidelberg, Germany, 2015; ISBN 978-3662448281.
11. Longo, A.; Verrucchi, R.; Aversa, L.; Tatti, R.; Ambrosio, A.; Orabona, E.; Coscia, U.; Carotenuto, G.; Maddalena, P. Graphene oxide prepared by graphene nanoplatelets and reduced by laser treatment. *Nanotecnology* **2017**, *28*, 224002. [CrossRef]
12. Sun, L.; Fugetsu, B. Mass production of graphene oxide from expanded graphite. *Mater. Lett.* **2013**, *109*, 207–210. [CrossRef]
13. Songfen, P.; Hui-Ming, C. The reduction of graphene oxide. *Carbon* **2012**, *50*, 3210.
14. Seung, H.H. *Thermal Reduction of Graphene Oxide—Physics and Applications of Graphene—Experiments*; Mikhailov, S., Ed.; InTech: Rijeka, Yugoslavia, 2011. [CrossRef]
15. Park, S.; An, J.; Potts, J.R.; Velamakanni, A.; Murali, S.; Ruoff, R.S. Hydrazine-reduction of graphite- and graphene oxide. *Carbon* **2011**, *49*, 3019–3023. [CrossRef]
16. Huang, L.; Liu, Y.; Ji, L.C.; Xie, Y.Q.; Wang, T.; Shi, W.Z. Pulsed laser assisted reduction of graphene oxide. *Carbon* **2011**, *49*, 2431–2436. [CrossRef]

17. Zhu, Y.; Murali, S.; Stoller, M.D.; Velamakanni, A.; Piner, R.D.; Ruoff, R.S. Microwave assisted exfoliation and reduction of graphite oxide for ultracapacitors. *Carbon* **2010**, *48*, 2118–2122. [CrossRef]
18. Hassan, H.M.A.; Abdelsayed, V.; Khder, A.E.R.S.; AbouZeid, K.M.; Terner, J.; El-Shall, M.S.; Al-Resayes, S.I.; El-Azhary, A.A. Microwave synthesis of graphene sheets supporting metal nanocrystals in aqueous and organic media. *J. Mater. Chem.* **2009**, *19*, 3832–3837. [CrossRef]
19. Furst, A.; Berlo, R.C.; Hooton, S. Hydrazine as a reducing agent for organic compounds (catalytic hydrazine reductions). *Chem. Rev.* **1965**, *65*, 51–68. [CrossRef]
20. Bo, Z.; Zhu, W.; Tu, X.; Yang, Y.; Mao, S.; He, Y.; Chen, J.; Yan, J.; Cen, K. Instantaneous reduction of graphene oxide paper for supercapacitor electrodes with unimpeded liquid permeation. *J. Phys. Chem. C* **2014**, *118*, 13493–13502. [CrossRef]
21. Abdelsayed, V.; Moussa, S.; Hassan, H.; Aluri, H.S.; Collinson, M.M.; El-Shall, M.S. Photothermal deoxygenation of graphite oxide with laser excitation in solution and graphene-aided increase in water temperature. *J. Phys. Chem. Lett.* **2010**, *1*, 2804–2809. [CrossRef]
22. Gao, W.; Singh, N.; Song, L.; Liu, Z.; Reddy, A.L.; Ci, L.; Vajtai, R.; Zhang, Q.; Wei, B.; Ajayan, P.M. Direct laser writing of micro-supercapacitors on hydrated graphite oxide films. *Nat. Nanotechnol.* **2011**, *6*, 496–500. [CrossRef] [PubMed]
23. Zhou, Y.; Bao, Q.; Varghese, B.; Tang, L.A.; Tan, C.K.; Sow, C.H.; Loh, K.P. Microstructuring of graphene oxide nanosheets using direct laser writing. *Adv. Mater.* **2010**, *22*, 67–71. [CrossRef]
24. Zhang, Y.; Guo, L.; Wei, S.; He, Y.; Xia, H.; Chen, Q.; Sun, H.-B.; Xiao, F.-S. Direct imprinting of microcircuits on graphene oxides film by femtosecond laser reduction. *Nano Today* **2010**, *5*, 15–20. [CrossRef]
25. Sokolov, D.A.; Shepperd, K.R.; Orlando, T.M. Formation of graphene features from direct laser-induced reduction of graphite oxide. *J. Phys. Chem. Lett.* **2010**, *1*, 2633–2636. [CrossRef]
26. Sokolov, D.A.; Rouleau, C.M.; Geohegan, D.B.; Orlando, T.M. Excimer laser reduction and patterning of graphite oxide. *Carbon* **2013**, *53*, 81–89. [CrossRef]
27. Orabona, E.; Ambrosio, A.; Longo, A.; Carotenuto, G.; Nicolais, L.; Maddalena, P. Holographic patterning of graphene-oxide films by light-driven reduction. *Opt. Lett.* **2014**, *39*, 4263–4266. [CrossRef] [PubMed]
28. Karrer, P. The Chemistry of Vitamins A and C. *Chem. Rev.* **1934**, *14*, 17–30. [CrossRef]
29. Loeffler, H.J.; Ponting, J.D. Ascorbic Acid. *Ind. Eng. Chem. Anal. Ed.* **1942**, *14*, 846–849. [CrossRef]
30. Austin, J.; Partridge, D.A. *Vitamin C: Its Chemistry and Biochemistry*; Davies, M.B., Ed.; Royal Society of Chemistry: London, UK, 1991.
31. Zhang, J.; Yang, H.; Sheng, G.; Cheng, P.; Zang, J.; Guo, S. Reduction of graphene oxide vial-ascorbic acid. *Chem. Commun.* **2010**, *46*, 1112–1114. [CrossRef]
32. Liu, J.; Liu, L.; Wu, X.; Zhang, X.; Li, T. Environmentally friendly synthesis of graphene–silver composites with surface-enhanced Raman scattering and antibacterial activity via reduction with l-ascorbic acid/water vapor. *Newj. Chem.* **2015**, *39*, 5272–5281. [CrossRef]
33. Habte, A.T.; Ayele, D.W. Synthesis and characterization of reduced graphene oxide (rgo) started from graphene oxide (go) using the tour method with different Parameters. *Adv. Mater. Sci. Eng.* **2019**, *15*, 1–9. [CrossRef]
34. Fernández-Merino, M.J.; Guardia, L.; Paredes, J.L.; Villar-Rodil, S.; Solís-Fernández, P.; Martínez-Alonso, A.; Tascón, J.M.D. Vitamin C is an ideal substitute for hydrazine in the reduction of graphene oxide suspensions. *J. Phys. Chem. C* **2010**, *214*, 6426–6432. [CrossRef]
35. Chen, D.; Li, L.; Guo, L. an environment-friendly preparation of reduced graphene oxide nanosheets via amino acid. *Nanotecnology* **2011**, *22*, 325601. [CrossRef] [PubMed]
36. Bose, S.; Kuila, T.; Mishra, A.K.; Kim, N.H.; Lee, J.H. Dual role of glycine as a chemical functionalize and a reducing agent in the preparation of graphene: An environmentally friendly method. *J. Mater. Chem.* **2012**, *22*, 9696–9703. [CrossRef]
37. Wang, Y.; Shi, Z.; Yin, J. Facile Synthesis of soluble graphene via a green reduction of graphene oxide in tea solution and its biocomposites. *Acs Appl. Mater. Interfaces* **2011**, *3*, 1127–1133. [CrossRef]
38. Costa, M.N.; Veigas, B.; Jacob, J.M.; Santos, D.S.; Gomes, J.; Baptista, P.V.; Martins, R.; Inácio, J.; Fortunato, E. A low cost, safe, disposable, rapid and self-sustainable paper-based platform for diagnostic testing: Lab-on-paper. *Nanotechnology* **2014**, *25*, 094006. [CrossRef]
39. Lua, Y.; Wang, H.; Zhao, W.; Zhang, M.; Qin, H.; Xie, Y. Flexible, stretchable sensors for wearable health monitoring: Sensing mechanisms, materials, fabrication strategies and features. *Sensor* **2018**, *18*, 645–652.

40. Akyazi, T.; Basabe-Desmonts, L.; Benito-Lopez, F. Review on microfluidic paper-based analytical devices towards commercialization. *Anal. Chim. Acta* **2017**, *1001*, 1–17. [CrossRef] [PubMed]

41. Song, S.; Zhai, Y.; Zhang, Y. Bioinspired graphene oxide/polymer nanocomposite paper with high strength, toughness, and dielectric constant. *Appl. Mater. Interfaces* **2016**, *8*, 31264–31272. [CrossRef] [PubMed]

42. Saha, B.; Baek, S.; Lee, J. Highly sensitive bendable and foldable paper sensors based on reduced graphene oxide. *Appl. Mater. Interfaces* **2017**, *9*, 4658–4666. [CrossRef]

43. Luong, N.D.; Pahimanolis, N.; Hippi, U.; Korhonen, J.T.; Ruokolainen, J.; Johansson, L.-S.; Namd, J.-D.; Seppälä, J. Graphene/cellulose nanocomposite paper with high electrical and mechanical performances. *J. Mater. Chem.* **2011**, *21*, 13991–13998. [CrossRef]

44. Faniyi, I.O.; Fasakin, O.; Olofnjana, B.; Adekunle, A.S.; Oluwasusi, T.V.; Eleruja, M.A.; Ajayi, E.O.B. The comparative analyses of reduced graphene oxide (RGO) prepared via green, mild and chemical approaches. *SN Appl. Sci.* **2019**, *1*, 1181. [CrossRef]

45. Buerger, M.J. *X-Ray Crystallography*; Wiley: New York, NY, USA, 1942.

46. Cullity, B.D. *Elements of X-Ray Diffraction, Reading*; Addison-Wesley: Boston, MA, USA, 1956.

47. Sharma, R.; Chadha, N.; Saini, P. Determination of defect density, crystallite size and number of graphene layers in graphene analogues using X-ray diffraction and Raman spectroscopy. *Indian J. Pure Appl. Phys.* **2017**, *55*, 625–629.

48. Dimotrov, A.S.; Kralchevsky, P.A.; Nicolov, A.D.; Noshi, H.; Matsumoto, M. Contact angle measurements with sessile drops and bubbles. *J. Colloid Interface Sci.* **1991**, *145*, 279–282. [CrossRef]

49. Kim, J.; Khoh, W.-H.; Weea, B.-H.; Hong, J.-D. Fabrication of flexible reduced graphene oxide–TiO2 freestanding films for supercapacitor application. *Rsc Adv.* **2015**, *5*, 9904–9911. [CrossRef]

50. Cho, Y.H.; Kim, H.W.; Lee, H.D.; Shin, J.E.; Yoo BM Park, H.B. Water and ion sorption, diffusion, and transport in graphene oxide membranes revisited. *J. Membr. Sci.* **2017**, *544*, 425–435. [CrossRef]

51. Chua, C.K.; Pumera, M. Chemical reduction of graphene oxide: A synthetic chemistry viewpoint. *Chem. Soc. Rev.* **2014**, *43*, 291–312. [CrossRef] [PubMed]

52. Gao, J.; Liu, F.; Liu, Y.; Ma, N.; Wang, Z.; Zhang, X. Environment-Friendly Method to Produce Graphene That Employs Vitamin C and Amino Acid. *Chem. Mater.* **2010**, *22*, 2213–2218. [CrossRef]

Article

Local Structure Analysis on Si-Containing DLC Films Based on the Measurement of C K-Edge and Si K-Edge X-ray Absorption Spectra

Kazuhiro Kanda [1,*], Shuto Suzuki [1], Masahito Niibe [1], Takayuki Hasegawa [1,2], Tsuneo Suzuki [3] and Hedetoshi Saitoh [3]

[1]   Laboratory of Advanced Science and Technology for Industry, University of Hyogo, 3-1-2 Koto, Kamigori-cho, Ako-gun, Hyogo 678-1205, Japan
[2]   Synchrotron Analysis L.L.C., 3-2-24 Wadamiya-dori, Hyogo-ku, Kobe 652-0884, Japan
[3]   Graduate School of Engineering, Nagaoka University of Technology, 16031-1 Kamitomioka, Nagaoka, Niigata 940-2188, Japan
*   Correspondence: kanda@lasti.u-hyogo.ac.jp; Tel.: +81-791-58-0476

Received: 2 March 2020; Accepted: 29 March 2020; Published: 30 March 2020

Abstract: In this paper, the local structure of silicon-containing diamond-like carbon (Si-DLC) films is discussed based on the measurement of C K-edge and Si K-edge near-edge x-ray absorption fine structure (NEXAFS) spectra using the synchrotron radiation of 11 types of Si-DLC film fabricated with various synthesis methods and having different elemental compositions. In the C K-edge NEXAFS spectra of the Si-DLC films, the σ* band shrunk and shifted to the lower-energy side, and the π* peak broadened with an increase in the Si content in the Si-DLC films. However, there were no significant changes observed in the Si K-edge NEXAFS spectra with an increase in the Si content. These results indicate that Si–Si bonding is not formed with precedence in Si-DLC film.

Keywords: Si-containing diamond-like carbon film; near-edge X-ray absorption fine structure; dependence on the elemental composition

---

1. Introduction

Diamond-like carbon (DLC) film is amorphous carbon (a-C) film, which has a disordered form of C consisting of sp^2 and sp^3 hybridized bonds [1,2]. DLC films are widely used as coating materials in various industrial fields due to their superior properties, such as their high hardness, low friction coefficient, chemical inertness, and gas barrier [3–5]. Over the last two decades, various novel DLC films have been synthesized for several industrial purposes. Hetero-atom-containing DLC films were developed because the existence of hetero-atoms in DLC film has the potential to improve the film properties. In particular, the incorporation of Si into DLC films has resulted in an increase in thermal stability [6,7] and resistance to oxidation [8,9], and a decrease in friction in a humid atmosphere [10–12] and intrinsic stresses [13,14]. Therefore, silicon-containing DLC (Si-DLC) films have attracted an increasing amount of attention in various industrial fields due to these properties [15–18].

Near-edge X–ray absorption fine structure (NEXAFS) spectroscopy using synchrotron radiation is known to be sensitive to the local structure around the absorber atom. In near-edge X–ray absorption fine structure (NEXAFS) spectroscopy, the out-coming current from the sample is recorded as a function of the photon excitation energy around the core level of the target element using synchrotron radiation. Transitions occurring via the unoccupied states of the x-ray absorbing elements, whose energy shifts are large, can be observed in this technique. Therefore, NEXAFS is known to be sensitive to the local structure around the absorber atom, in comparison with other spectral evaluation methods. It also does not require long-range ordering and is sensitive to the local bonding structure. Therefore, NEXAFS spectroscopy is an adequate method for evaluating DLC films, which have an amorphous structure. The NEXAFS spectra of Si-DLC films have also been measured, e.g., C K-edge NEXAFS spectra [19–23] and Si K-edge NEXAFS spectra [24,25]. We previously reported a comprehensive analysis on Si-DLC films based on the measurements of C K-edge NEXAFS spectra [26]. In the present study, we measured the C K-edge NEXAFS spectra and Si K-edge NEXAFS spectra of 11 types of Si-DLC film fabricated using various synthesis methods. The elementary analysis was conducted using a combination of Rutherford backscattering spectrometry (RBS) and elastic recoil detection analysis (ERDA). We discuss the composition ratio of Si dependency on the local structure and the chemical environments of C and Si in Si-DLC films.

2. Materials and Methods

We analyzed 11 types of Si-DLC film, which were provided by enterprises, public organizations, and universities, by measuring the C K-, Si K-, and Si L-edge NEXAFS spectra. These DLC samples were collected by the DLC research group in Japan, which has been working on the classification of DLC films [27]. The Si-DLC films were deposited on Si substrates by using various methods, such as physical vapor deposition (PVD) and chemical vapor deposition (CVD). The desired film thickness was 200 nm.

The RBS and ERDA measurements were performed using an electrostatic accelerator (NT-1700HS: Nisshin-High Voltage Co., Kyoto, Japan) located at the Extreme Energy-Density Research Institute, Nagaoka University of Technology, Nagaoka, Japan. Details of RBS and ERDA measurements are described in refs. [28–30]. The He$^+$ ions accelerated to 2.5 MeV were used as the incident beam at 72° with respect to the normal surface of a sample. In the RBS measurement, a small fraction—~0.1%—of high-energy He$^+$ ions elastically scattered by the sample were captured with a solid-state detector (SSD) arranged at 12° with respect to the normal surface of a sample toward the incident beam. RBS was applied to determine the atomic fractions of C and Si in the samples. In the ERDA measurement, the He$^+$ ions elastically collided with H atoms in a sample. The H atoms that were ejected from a sample were detected with an SSD placed at 78° to the normal surface of a sample in the opposite direction to the incident beam. RBS and ERDA signals were simultaneously detected and ERDA was applied to determine the atomic fraction of H in the samples. The estimation error of the RBS/ERDA measurement was 0.5 at.%, because the determination of the content was simulated in 1 at.% steps. The obtained elemental compositions of these samples are listed in Table 1. The ratios of Si/(C + Si) in these films ranged from 0.03 to 0.39.

Table 1. Elemental composition of silicon-containing diamond-like carbon (Si-DLC) films obtained from Rutherford backscattering spectrometry (RBS)/elastic recoil detection analysis (ERDA) measurements and the $sp^2/(sp^2 + sp^3)$ ratio estimated from C K-edge near-edge x-ray absorption fine structure (NEXAFS).

Sample	H Content (at.%)	Si Content (at.%)	Si/(C + Si)	$sp^2/(sp^2 + sp^3)$
1	23	2	0.03	0.61
2	22	5	0.07	0.64
3	20	7	0.09	0.55
4	19	7	0.09	0.31
5	22	8	0.11	0.58
6	29	12	0.17	0.50
7	26	15	0.20	0.44
8	27	15	0.21	0.50
9	41	14	0.23	0.16
10	20	18	0.23	0.63
11	33	26	0.39	0.37

NEXAFS measurements were carried out at the NewSUBARU synchrotron facility of the University of Hyogo [31]. The Si K-edge NEXAFS spectra were measured at beamline 05A, where synchrotron radiation provided by a bending magnet was dispersed using a double crystal monochromator (DXM) [32]. The Si K-edge NEXAFS spectra were measured in the energy range from 1830 to 1880 eV using InSb (111) as a dispersive crystal. The Si L- and C K-edge NEXAFS spectra were measured at beamline 09A, where synchrotron radiation provided by an 11-m undulator was dispersed using a varied spacing planer-grating monochromator [33–35]. The Si L-edge and C K-edge NEXAFS spectra were measured in the energy range from 90 to 120 eV and from 275 to 335 eV, respectively. At both beamlines, monochromatized synchrotron radiation was irradiated on the sample at 54.7° (magic angle) with respect to the normal surface of a sample. All NEXAFS spectra were measured with the total electron yield (TEY) mode. The electrons coming from the sample were detected by measuring the current from the earth to the sample, I_s. The intensity of the incident photon beams, I_0, was measured by monitoring the photocurrent from a gold mesh, which was placed in front of the sample. The absorption signal was obtained by the ratio between the out-going electron intensity from the sample, I_s, and I_0. In order to support the correctness of the measurements, all NEXAFS spectra of each sample were recorded at a minimum of three locations.

3. Results and Discussion

Figure 1 shows the measured C K-edge NEXAFS spectra of the 11 Si-DLC film samples. The spectra are displayed from the highest (top) to lowest (bottom) Si/(C + Si) ratio. Figure 2 shows the C K-edge NEXAFS spectrum of the Si-DLC film that had the highest content of Si (sample 11), along with those of diamond powder, HOPG, β-SiC powder, and a typical commercial DLC film that was deposited on a 200 nm-thick Si wafer by using the ion plating method. The sharp π^* peak observed at 285.38 eV in the C K-edge NEXAFS spectra of HOPG is ascribed to the C1s→π^* resonance transition originating from the C=C bonds [36]. The peak of the C1s→π^* resonance transition was observed at 284.4 eV in the typical DLC film. This peak was not observed in the C K-edge NEXAFS spectra of diamond powder and β-SiC powder. Structural peaks were observed in the 285–310 eV region of the C K-edge spectra of HOPG, diamond powder, and β-SiC powder. The peaks in the spectra of diamond powder and β-SiC powder are ascribed to the C1s→σ^* resonance transition generated from the C–C bonds and the C–Si bonds, respectively [36–38]. The peaks in the spectra of HOPG are ascribed to the C1s→σ^* resonance transition generated from the C–C bonds and C=C bonds. In the C K-edge spectra of the DLC film and Si-DLC film, the C1s→σ^* resonance transition generated from the C–C bonds, C=C bonds, or C–Si bonds was observed as a broad band at about 285–320 eV, because DLC and Si-DLC have amorphous structures.

The spectral profiles of the C K-edge NEXAFS spectra of the Si-DLC films were reported to change along with the ratio of Si/(C + Si) in the films [26]. As the Si/(C + Si) ratio increases in the film, 1) the σ^* band shrinks due to the change from σ(C–C) to σ(C–Si) and 2) broadening of the π^* peak can be seen due to an increase in the composition ratio of C=C–Si sites.

The $sp^2/(sp^2 + sp^3)$ ratio of carbon atoms in DLC films is the most important information for understanding the properties of such films. This ratio depends on many factors, such as the deposition method and deposition condition. The $sp^2/(sp^2 + sp^3)$ ratio decreases with an increasing H content in DLC film [39]. The absolute $sp^2/(sp^2 + sp^3)$ ratio of C atoms in DLC film can be determined with a high accuracy because the 1s→π^* resonance transition can be separately observed in the C K-edge NEXAFS spectrum [39–41]. The amount of sp^2 hybridized C atoms can be extracted by normalizing the area of resonance corresponding to the 1s→π^* transitions at 285.4 eV with a large section of the spectrum. The absolute $sp^2/(sp^2 + sp^3)$ ratio was determined by comparing it with that from the NEXAFS spectrum of HOPG, which can be regarded as the full construction of sp^2 hybridized C. The $sp^2/(sp^2 + sp^3)$ ratios estimated from the C-K edge NEXAFS spectra of the Si-DLC films are summarized in Table 1. These values are the average $sp^2/(sp^2 + sp^3)$ ratio obtained from C-K edge NEXAFS spectra measured at three locations of each sample. It was reported that the $sp^2/(sp^2 + sp^3)$ ratios of the Si-DLC films gradually decreased with an increasing Si/(C + Si) ratio in our previous study [26]. It can be considered that Si atoms only coupled to C atoms by single bonding in Si-DLC films. The present results are almost in accordance with this tendency; however, several exceptions, for example, sample 10, were obtained. The $sp^2/(sp^2 + sp^3)$ ratios of carbon films depend on the deposition method, deposition condition, density, and hydrogen content. In sample 10, the $sp^2/(sp^2 + sp^3)$ ratio was considered to be lowered by its high content of hydrogen.

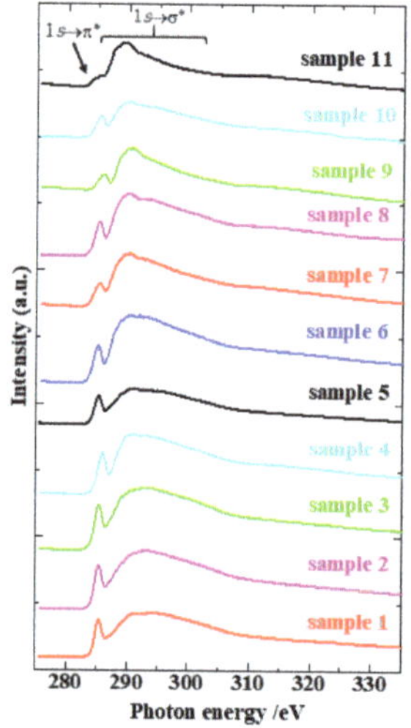

Figure 1. C *K*-edge NEXAFS spectra of 11 types of Si-DLC film samples. Spectra are displayed from the highest (top) to lowest (bottom) Si/(C + Si) ratio.

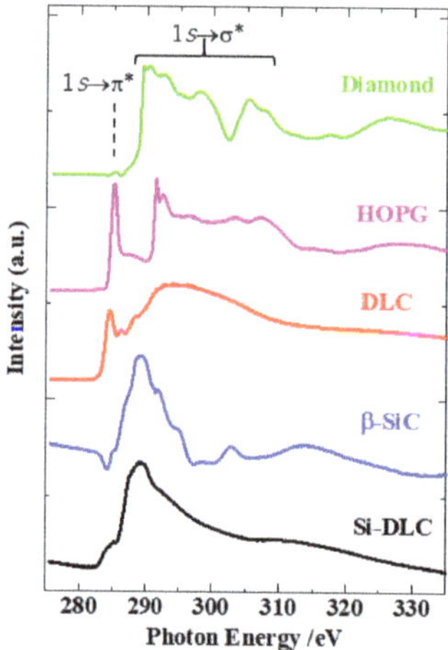

Figure 2. C *K*-edge NEXAFS spectra of Si-DLC film (sample 11), diamond powder, HOPG, typical DLC film, and β–SiC powder.

Figure 3 shows the Si *K*-edge NEXAFS spectra of the 11 Si-DLC film samples. The spectra are displayed in the same order as in Figure 1. Figure 4 shows the Si *K*-edge spectra of sample 11, along with those of SiO_2 powder, β-SiC powder, a-Si:H film, and Si wafer. The energy of the first inflection point, E_0, and that of the white line, representing an electronic Si $1s{\rightarrow}t_2$ transition, E_{max}, depend on the electronegativity of the atoms surrounding the Si [38]. These energies shift towards a higher energy with an increasing positive charge on the absorber. An intense sharp peak was observed at 1846.8 eV in the spectrum of SiO_2 powder [42]. The E_0 and E_{max} of the Si-DLC film were slightly higher than those of the Si wafer and a-Si:H film, much lower than those of SiO_2 powder, and similar to those of SiC, as reported by V. Palshin et al. [25]. As shown in Figure 3, E_0, E_{max}, and other spectral features did not vary systematically along with the Si/(C + Si) ratio in the present study.

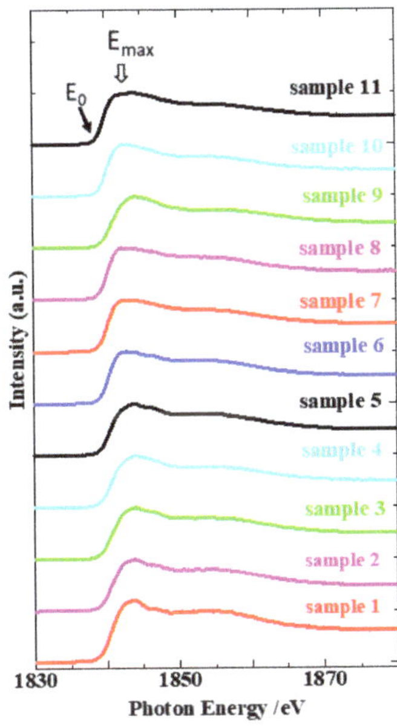

Figure 3. Si K-edge NEXAFS spectra of 11 types of Si-DLC film samples. Spectra are displayed from the highest (top) to lowest (bottom) Si/(C + Si) ratio. A filled arrow and open arrow indicate the position of E_0 and E_{max} of the Si $1s \rightarrow t_2$ transition, respectively.

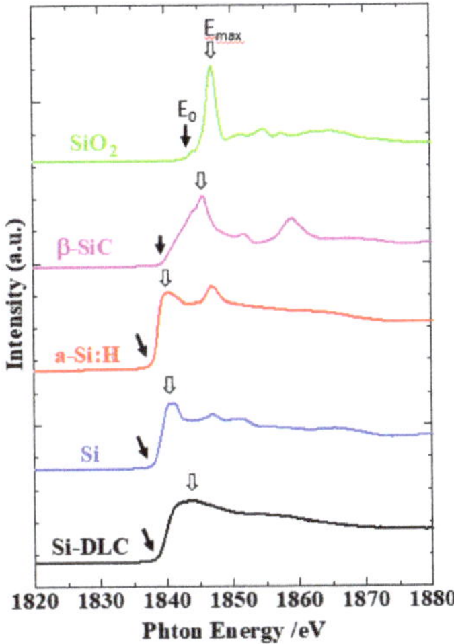

Figure 4. Si K-edge NEXAFS spectra of Si-DLC film (sample 11), SiO_2 powder, β–SiC powder, a-Si:H film, and Si wafer. Filled arrows and open arrows indicate the position of E_0 and E_{max} of the Si $1s \rightarrow t_2$ transition, respectively.

Figure 5 shows the Si *L*-edge NEXAFS spectra of the 11 Si-DLC film samples. Figure 6 shows the Si *L*-edge NEXAFS spectra of sample 11, along with those of SiO_2 powder, β-SiC powder, a-Si:H film, and Si wafer. In the spectrum of SiO_2, characteristic peaks were observed at 105.4, 106.0, and 107.8 eV, which were assigned to the transitions $2p_{3/2} \rightarrow a_1$, $2p_{1/2} \rightarrow a_1$, and $2p \rightarrow t$, respectively [43]. These characteristic peaks derived from oxidized Si were observed in the Si *L*-edge spectra of samples 3 and 10, as shown in Figure 5, and those of a-Si:H film and Si wafer in Figure 6. These samples were obviously oxidized naturally. However, the characteristic peak derived from oxidized Si, which was observed at 1846.8 eV in the Si *K*-edge spectrum of SiO_2 powder, was not observed in the Si *K*-edge NEXAFS spectra of the above samples. This discordance is ascribable to the difference in the detection depth of Si *K*-edge, with a value of ~1850 eV, and Si *L*-edge, with a value of ~100 eV. Natural oxidation of these samples was considered to occur in the neighborhood of the surfaces of each sample. The energy of the first inflection point, E_0, of *L*-edge elemental Si is 100 eV and that of silicon oxide is 105 eV [43]. The E_0 of *L*-edge of the Si-DLC films, which are not oxidized, is ≈104 eV. This resembled that of SiC, is higher than those of Si wafer and a-Si:H film, and is lower than that of SiO_2 powder. The relation of E_0 positions due to the materials is similar to that in the Si *K*-edge.

As described above, the chemical environment of C atoms varied along with the Si/(C + Si) ratio in the Si-DLC film. However, that of Si atoms was not affected by this ratio. These results indicate that Si atoms in the Si-DLC film dominantly bonded to C atoms in the Si/(C + Si) region, with a value of less than 0.39. In other words, Si–Si bonding did not occur in this Si/(C + Si) ratio region. Palshin et. al. concluded that Si atoms in Si-DLC films are surrounded by four C atoms, as revealed by Si *K*-edge extended x-ray absorption fine structure (EXAFS) measurement [25]. The measurements in the present study agree with their conclusion. However, the Si/(C + Si) ratio of the Si-DLC films they used was 0.125, but that of sample 11 in the present study was ≈0.4. In Si-DLC film with an Si/(C + Si) ratio of 0.39, the DLC structure cannot be constructed without coupling of the Si–Si bond. Therefore, the Si *K*- and *L*-edge NEXAFS spectra do not depend on the Si/(C + Si) ratio in Si-DLC films. Additionally, not only do Si atoms tend to enter between C atoms, but the chemical environment of an Si atom coupled to a C atom is close to that of an Si atom coupled to Si atoms, as shown in Figure 4.

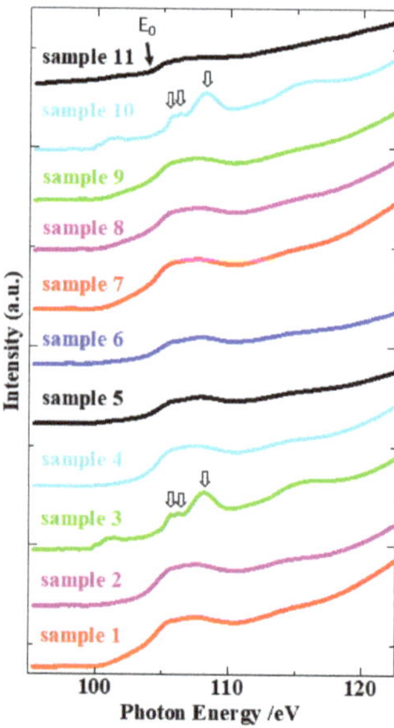

Figure 5. Si *L*-edge NEXAFS spectra of 11 types of Si-DLC film samples. Spectra are displayed from the highest (top) to lowest (bottom) Si/(C + Si) ratio.

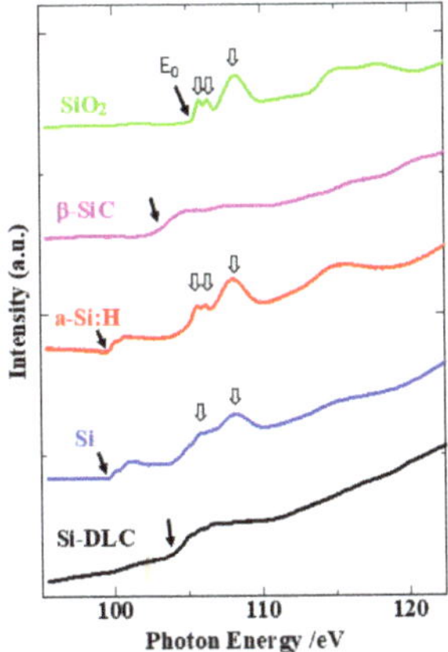

Figure 6. Si *L*-edge NEXAFS spectra of Si-DLC film (sample 11), SiO$_2$ powder, β–SiC powder, a-Si:H film, and Si wafer. Filled arrows indicate the position of E_0. Open arrows indicate the peaks derived from the transitions of the oxidized Si.

4. Conclusions

In this study, we investigated the local structure of Si-DLC films by measuring the C *K*-edge and Si *K*-edge NEXAFS spectra of 11 types of Si-DLC film with Si/(C + Si) ratios ranging from 0.03 to 0.39. The spectral features of the C *K*-edge NEXAFS spectra of the Si-DLC films varied along with the Si/(C + Si) ratio: 1) the σ* band at 275–335 eV shrank and 2) the π peak at 285 eV broadened. These spectral changes are ascribable to the change from σ(C–C) to σ(C–Si) and an increase in the composition ratio of C=C–Si sites with the Si/(C+Si) ratio in the Si-DLC films. However, the spectral feature of the Si *K*-edge NEXAFS spectra of the Si-DLC films did not significantly change along with the Si/(C + Si) ratio. This is considered due to the fact that the Si atom positioned between C atoms in Si-DLC films and/or the chemical environment of an Si atom does not change much between Si–C and Si–Si bonding. Namely, the electronic state of Si in the Si-DLC film is not affected by the Si/(C + Si) ratio.

Author Contributions: Conceptualization, K.K. and S.S.; investigation, K.K., S.S., M.N., T.H., and T.S.; writing—original draft preparation, S.S.; writing—review and editing, K.K.; supervision, H.S.; project administration, K.K.; funding acquisition, K.K. All authors have read and agreed to the published version of the manuscript.

Funding: This research was funded by JSPS KAKENHI, Grant No. JP17K06799, and a Research Grant from the Nippon Sheet Glass Foundation for Materials Science and Engineering.

Acknowledgments: The authors would like to thank Akira Heya of the University of Hyogo for providing the a-Si:H film used as a reference material. We also thank the NewSUBARU staff for their efforts regarding the stable operation of the NewSUBARU ring.

Conflicts of Interest: The authors declare no conflicts of interest. The funders had no role in the design of the study; in the collection, analyses, or interpretation of data; in the writing of the manuscript; or in the decision to publish the results.

References

1. Aisenberg, S.; Chabot, R. Ion-beam deposition of thin films of diamondlike carbon. *J. Appl. Phys.* **1971**, *42*, 2953–2958. [CrossRef]
2. Robertson, J. Properties of diamond-like carbon. *J. Surf. Coat. Technol.* **1992**, *50*, 185–203. [CrossRef]
3. Aisenberg, S. Properties and applications of diamondlike carbon films. *J. Vac. Sci. Technol. A* **1984**, *2*, 369–371. [CrossRef]
4. Grill, A. Diamond-like carbon: State of the art. *Diam. Relat. Mater.* **1999**, *8*, 428–434. [CrossRef]
5. Robertson, J. Diamond-like amorphous carbon. *Mater. Sci. Eng. R* **2002**, *37*, 129–281. [CrossRef]

6. Camargo, S.S., Jr.; Santos, R.A.; Baia-Neto, A.L.; Carius, R.; Finger, F. Structural modifications and temperature stability of silicon incorporated diamond-like a-C:H films. *Thin Solid Films* **1998**, *332*, 130–135. [CrossRef]

7. Wu, W.J.; Hon, M.H. Thermal stability of diamond-like carbon films with added silicon. *Surf. Coat. Technol.* **1999**, *111*, 134–140. [CrossRef]

8. Camargo, S.S., Jr.; Baia-Neto, A.L.; Santos, R.A.; Freire, F.L.; Carius, R.; Finger, F. Improved high-temperature stability of Si incorporated a-C:H films. *Diam. Relat. Mater.* **1998**, *7*, 1155–1162. [CrossRef]

9. Kidena, K.; Endo, M.; Takamatsu, H.; Imai, R.; Niibe, M.; Yokota, K.; Tagawa, M.; Furuyama, Y.; Komatsu, K.; Saitoh, H.; et al. Hyperthermal atomic oxygen beam irradiation effect on the hydrogenated Si-doped DLC film. *Trans. Mater. Res. Soc. Jpn.* **2015**, *40*, 353–358. [CrossRef]

10. Oguri, K.; Arai, T. Tribological properties and characterization of diamond-like carbon coatings with silicon prepared by plasma-assisted chemical vapour deposition. *Surf. Coat. Technol.* **1991**, *47*, 710–721. [CrossRef]

11. Kim, M.-G.; Lee, K.-R.; Eun, K.Y. Tribological behavior of silicon-incorporated diamond-like carbon films. *Surf. Coat. Technol.* **1999**, *112*, 204–209. [CrossRef]

12. Chen, T.; Wu, X.; Ge, Z.; Ruan, J.; Lv, B.; Zhang, J. Achieving low friction and wear under various humidity conditions by co-doping nitrogen and silicon into diamond-like carbon films. *Thin Solid Films* **2017**, *638*, 375–382. [CrossRef]

13. Damasceno, J.C.; Camargo, S.S.; Freire, F.L.; Carius, R. Deposition of Si-DLC films with high hardness, low stress and high deposition rates. *Surf. Coat. Technol.* **2000**, *133*, 247–252. [CrossRef]

14. Lee, C.S.; Lee, K.R.; Eun, K.Y.; Yoon, K.H.; Han, J.H. Structure and properties of Si incorporated tetrahedral amorphous carbon films prepared by hybrid filtered vacuum arc process. *Diam. Relat. Mater.* **2002**, *11*, 198–203.

15. Lee, K.-R.; Kim, M.-G.; Cho, S.-J.; Eun, K.Y.; Seong, T.-Y. Structural dependence of mechanical properties of Si incorporated diamond-like carbon films deposited by RF plasma-assisted chemical vapour deposition. *Thin Solid Films* **1997**, *308–309*, 263–267. [CrossRef]

16. Okpalugo, T.I.T.; Ogwu, A.A.; Maguire, P.D.; McLaughlin, J.A.D. Platelet adhesion on silicon modified hydrogenated amorphous carbon films. *Biomaterials* **2004**, *25*, 239–245. [CrossRef]

17. Kalin, M.; Oblak, E.; Akbari, S. Evolution of the nano-scale mechanical properties of tribofilms formed from low- and high-SAPS oils and ZDDP on DLC coatings and steel. *Tribol. Int.* **2016**, *96*, 43–56. [CrossRef]

18. Bociaga, D.; Sobczyk-Guzenda, A.; Szymanski, W.; Jedrzejczak, A.; Jastrzebska, A.; Olejnik, A.; Jastrzebski, K. Mechanical properties, chemical analysis and evaluation of antimicrobial response of Si-DLC coatings fabricated on AISI 316 LVM substrate by a multi-target DC-RF magnetron sputtering method for potential biomedical applications. *Appl. Surf. Sci.* **2017**, *417*, 23–33. [CrossRef]

19. Monteiro, O.R.; Delplancke-Ogletree, M.-P. *Investigation of Non-Hydrogenated DLC: Si Prepared by Cathodic Arc*; LBNL-49944; University of California: Oakland, CA, USA, 2002.

20. Jung, H.-S.; Park, H.-H. Determination of bonding structure of Si, Ge, and N incorporated amorphous carbon films by near-edge x-ray absorption fine structure and ultraviolet Raman spectroscopy. *J. Appl. Phys.* **2004**, *96*, 1013–1018. [CrossRef]

21. Ray, S.C.; Bao, C.W.; Tsai, H.M.; Chiou, J.W.; Jan, J.C.; Krishna Kumar, K.P.; Pong, W.F.; Tsai, M.-H.; Wang, W.-J.; Hsu, C.-J.; et al. Electronic structure and bonding properties of Si-doped hydrogenated amorphous carbon films. *Appl. Phys. Lett.* **2004**, *85*, 4022–4024. [CrossRef]

22. Ray, S.C.; Okpalugo, T.I.T.; Papakonstantinou, P.; Bao, C.W.; Tsai, H.M.; Chiou, J.W.; Jan, J.C.; Pong, W.F.; McLaughlin, J.A.; Wang, W.J. Electronic structure and hardening mechanism of Si-doped/undoped diamond-like carbon films. *Thin Solid Films* **2005**, *482*, 242–247. [CrossRef]

23. Wada, A.; Ogaki, T.; Niibe, M.; Tagawa, M.; Saitoh, H.; Kanda, K.; Ito, H. Local structural analysis of a-SiC$_x$:H films formed by decomposition of tetramethylsilane in microwave discharge flow of Ar. *Diam. Relat. Mater.* **2011**, *20*, 364–367. [CrossRef]

24. Mastelaro, V.; Flank, A.M.; Fantini, M.C.A.; Bittencourt, D.R.S.; Carreño, M.N.P.; Pereyra, I. On the structural properties of a -Si$_{1-x}$C$_x$:H thin films. *J. Appl. Phys.* **1996**, *79*, 1324–1329. [CrossRef]

25. Palshin, V.; Tittsworth, R.C.; Fountzoulas, C.G.; Meletis, E.I. X-ray absorption spectroscopy, simulation and modeling of Si-DLC films. *J. Mater. Sci.* **2002**, *37*, 1535–1539. [CrossRef]

26. Kanda, K.; Niibe, M.; Wada, A.; Ito, H.; Suzuki, T.; Ohana, T.; Ohtake, N.; Saitoh, H. Comprehensive classification of near-edge X-ray absorption fine structure spectra of Si-containing diamond-like carbon thin films. *Jpn. J. Appl. Phys.* **2013**, *52*, 95504. [CrossRef]

27. Saitoh:, H. Current status on classification work of diamond-like carbon. In Proceedings of the 5th International Conference on New Diamond and Nano Carbons, Matsue, Japan, 16–20 May 2011.

28. Ohkawara, Y.; Ohshio, S.; Suzuki, T.; Ito, H.; Yatsui, K.; Saitoh, H. Dehydrogenation of nitrogen-containing carbon films by high-energy He^{2+} irradiation. *Jpn. J. Appl. Phys.* **2001**, *40*, 3359–3363. [CrossRef]

29. Ohkawara, Y.; Ohshio, S.; Suzuki, T.; Ito, H.; Yatsui, K.; Saitoh, H. Quantitative analysis of hydrogen in amorphous films of hydrogenated carbon nitride. *Jpn. J. Appl. Phys.* **2001**, *40*, 7007–7012. [CrossRef]

30. Igaki, J.; Saikubo, A.; Kometani, R.; Kanda, K.; Suzuki, T.; Niihara, K.; Matsui, S. Elementary analysis of diamond-like carbon film formed by focused-ion-beam chemical vapor deposition. *Jpn. J. Appl. Phys.* **2007**, *46*, 8003–8004. [CrossRef]

31. Ando, A.; Amano, S.; Hashimoto, S.; Kinosita, H.; Miyamoto, S.; Mochizuki, T.; Niibe, M.; Shoji, Y.; Terasawa, M.; Watanabe, T. VUV and soft X-ray light source "New SUBARU". In Proceedings of the IEEE Particle Accelerator Conference, Vancouver, BC, Canada, 12–16 May 1997; pp. 757–759.

32. Kanda, K.; Hasegawa, T.; Uemura, M.; Niibe, M.; Haruyama, Y.; Motoyama, M.; Amemiya, K.; Fukushima, S.; Ohta, T. Construction of a wide-range high-resolution beamline BL05 in NewSUBARU for X-ray spectroscopic analysis on industrial materials. *J. Phys. Conf. Ser.* **2013**, *425*, 132005. [CrossRef]

33. Niibe, M.; Mukai, M.; Kimura, H.; Shoji, Y. Polarization property measurement of the long undulator radiation using Cr/C multilayer polarization elements. *AIP Conf. Proc.* **2004**, *705*, 243–246.

34. Niibe, M.; Mukai, M.; Miyamoto, S.; Shoji, Y.; Hashimoto, S.; Ando, A.; Tanaka, T.; Miyai, M.; Kitamura, H. Characterization of Light Radiated from 11m Long Undulator. *AIP Conf. Proc.* **2004**, *705*, 576–579.

35. Kanda, K.; Okada, M.; Kang, Y.; Niibe, M.; Wada, A.; Ito, H.; Suzuki, T.; Matsui, S. Structural changes in diamond-like carbon films fabricated by Ga focused-ion-beam-assisted deposition caused by annealing. *Jpn. J. Appl. Phys.* **2010**, *49*. [CrossRef]

36. Batson, P.E. Carbon 1s near-edge-absorption fine structure in graphite. *Phys. Rev. B* **1993**, *48*, 2608–2610. [CrossRef] [PubMed]

37. Nithianandam, J.; Rife, J.C.; Windischmann, H. Carbon K edge spectroscopy of internal interface and defect states of chemical vapor deposited diamond films. *Appl. Phys. Lett.* **1992**, *60*, 135–137. [CrossRef]

38. Franke, R.; Bender, S.; Jüngerermann, H.; Kroscel, M.; Jansen, M. The determination of structural units in amorphous Si–B–N–C ceramics by means of Si, B, N and C K–XANES spectroscopy. *J. Electron Spectrosc. Relat. Phenom.* **1999**, *101–103*, 641–645. [CrossRef]

39. Saikubo, A.; Yamada, N.; Kanda, K.; Matsui, S.; Suzuki, T.; Niihara, N.; Saitoh, H. Comprehensive classification of DLC films formed by various methods using NEXAFS measurement. *Diam. Relat. Mater.* **2008**, *17*, 1743–1745. [CrossRef]

40. Kanda, K.; Kitagawa, T.; Shimizugawa, Y.; Haruyama, Y.; Matsui, S.; Terasawa, M.; Tsubakino, H.; Yamada, I.; Gejo, T.; Kamada, M. Characterization of hard DLC films formed by Ar gas cluster ion beam-assisted fullerene deposition. *Jpn. J. Appl. Phys.* **2002**, *41*, 4295–4298. [CrossRef]

41. Saikubo, A.; Kanda, K.; Niibe, M.; Matsui, S. Near-edge X-ray absorption fine-structure characterization of diamond-like carbon thin films formed by various method. *New Diam. Front. Carbon Technol.* **2006**, *16*, 235–244.

42. Li, D.; Bancroft, G.M.; Kasrai, M.; Fleet, M.E.; Secco, R.A.; Feng, X.H.; Tan, K.H.; Yang, B.X. X-ray absorption spectroscopy of silicon dioxide (SiO2) polymorphs: The structural characterization of opal. *Am. Mineral.* **1994**, *79*, 622–632.

43. Sammynaiken, R.; Naftel, S.; Sham, T.K.; Cheah, K.W.; Averboukh, B.; Huber, R.; Shen, Y.R.; Qin, G.G.; Ma, Z.C.; Zong, W.H. Structure and electronic properties of SiO$_2$/Si multilayer superlattices: Si K edge and $L_{3,2}$ edge X-ray absorption fine structure study. *J. Appl. Phys.* **2002**, *92*, 3000–3006. [CrossRef]

MDPI
St. Alban-Anlage 66
4052 Basel
Switzerland
Tel. +41 61 683 77 34
Fax +41 61 302 89 18
www.mdpi.com

Coatings Editorial Office
E-mail: coatings@mdpi.com
www.mdpi.com/journal/coatings